Gold on the Dunstan

Gold on the Dunstan

John McCraw

Square One Press
Book & Magazine Publishers
A Division of Mediaprint Services Limited.

Published by Square One Press, P. O. Box 2143, Dunedin, New Zealand. Phone 03 455-3117, Fax: 03 456 1053
Email: treeves@es.co.nz

ISBN 0 908562 59 4

Internet catalogue:
The Book Company of New Zealand:
http://www.book.co.nz

Book Design, Trevor Reeves.
Cover Picture, Judith Wolfe.
Stamping Battery, Mahinerangi
(arts site: www.arts.org.nz)

Produced by Mediaprint Services Limited,
P. O. Box 2143, Dunedin, New Zealand.
Printed by Otago University Print, Dunedin.

ISBN 0 908562 59 4

CONTENTS

PREFACE

This book is the third in a series recording historical events in the Clyde-Alexandra district of Central Otago. As with the earlier volumes it consists of a series of essays dealing with early gold mining, coal mining, and battles between farmers and miners over the use of land and water. The book covers more particularly the outlying districts, ranging from Campbell Creek, a tributary of the Waikaia River, through the Fraser Basin and Little Valley to Waikerikeri Valley in the foothills of the Dunstan Mountains.

Thousands of unknown miners worked on the goldfields. The only record of their presence is a name and a broad locality written on one of scores of butts in a book of Miners Rights stored along with hundreds of similar books in some corner of New Zealand Archives. With a Miner's Right a miner could not only prospect for gold, but if he found a payable prospect, he could peg out a claim 30 x 30 feet (9 x 9 m) without any further notification to officialdom.

It was only if he applied to the Warden's Court for 'Protection' for his claim while he was absent, that his name would be recorded in the Court's Application Book. If he applied for a water race from a stream or for a larger 'Extended claim,' then he would appear at the court for a 'hearing,' especially if someone put in an objection to his application, and his name would be recorded for posterity.

A problem facing the goldfields historian is the widespread use of local, ephemeral names for landscape features. Within a few of kilometres of Alexandra were places such as Solomons Face, Edwards Gully, Pensioner Point, Tucker Hill, Dunns Gully, Prospectors Point, Frenchmans Point, Italian Bend, The Bluffs, Sandy Hook and so on. These names were happily accepted by the Warden's Court but they seldom appeared on old maps or on plans of claims and are entirely absent from modern maps. This makes it very difficult to locate the position of claims or dams or the route of water races. But careful reading of newspapers can give a clue. For instance the exact location of the Campbell's Gully diggings was in doubt until a report was found which described the route over the Old Man Range. It mentioned that at a certain point, tents situated on the terrace above the diggings at Campbell Creek, first came into view. This viewpoint can still be identified and from it the site of main diggings at Campbell Creek can be deduced.

Archaeologists working for the Department of Conservation are doing a sterling job in identifying former settlements and buildings. The identification and excavation of old hotels and other buildings, including miners' huts are giving a much better idea of the way of life on the goldfields. Even the locations of complete villages have been confirmed. The exact position of the packers' village of Chamonix in the valley of Gorge Creek has always been in doubt. Recently archaeologists have discovered parallel rows of shallow excavations, which fit the description of a single street with a row of tents or huts on either side.

Although it is important that museums preserve smaller hardware relics, and particularly documents and photographs, it should always be remembered that the real goldfields museum is outdoors. Luckily there is growing interest in preserving goldfields relics in the field. Fortunately, many structures were built of stone and have lasted well. The dry climate of most of Central Otago has greatly helped to preserve relics and artefacts so that even the remains of sun-dried brick buildings are abundant. Until recently there was not the pressure, even in the towns, to destroy old buildings and structures in the name of progress. But this is changing. The incoming of large national retail outlets requiring huge amounts of space has seen old and historic buildings bulldozed, while the proliferation of rural 'lifestyle blocks' and the spread of vineyards is endangering water races, old mining dams, sites of old buildings, and so on.

As a first step all these historic features need to be recorded, even though at this stage little may be known about their origin. Because most are on private property it is important that the significance of surviving relics be explained to the landholders. In almost all cases they are interested to know the history of relics and will take pride in their preservation. They may also be interested in allowing access to interested visitors. Then plaques can be erected to explain the importance of the relics, followed by brochures giving locations and means of access. Provision for car parking on roadsides, gates or styles over fences and perhaps viewing points should also be thought about.

There are few places in New Zealand where the activities of the pioneers are so well preserved. Central Otago has a unique opportunity, not to 'exploit' them, but to show them to interested tourists without the razzmatazz that seems to often accompany historic tourism.

Comparison of Currency

Sums of money mentioned throughout this book have little relevance to modern values. One method of comparison is to convert the old value to the equivalent weight in gold. As gold was worth £3 17s. 6d. (£3.875) an ounce for much of the latter half of the Nineteenth Century, division of the sum of money by 3.875 will give its worth in ounces of gold. Multiplying the result by 600 (the value of an ounce of gold at present is aproximately $NZ600) gives some indication of the equivalent in today's currency. Using this method, comparative values are:

Old Currency	**Modern Currency (in NZ dollars)**
1s. (one shilling)	$8
2s. 6d (half a crown)	$20
£1 (one pound= 20 shillings)	$160
£3 17s. 6d. (value of one ounce of gold)	$600
£100	$16,000
£1,000	£160,000

Comparison of Weights

The **Troy** system of weight was used for weighing gold and the **Avoirdupois** system for weighing general goods

Troy		**Metric**
1 grain		= 0.064 grams
24 grains (gr)	= 1 pennyweight (dwt)	= 1.5 grams
20 pennyweights	= 1 ounce (oz)	= 31.1 grams
12 ounces	= 1 pound (lb)	= 373.2 grams

Avoirdupois		
1 ounce (oz)		= 28.35 grams
16 ounces	= 1 pound (lb)	= 453.6 grams
112 lbs	= 1 hundredweight (cwt)	= 50.8 kilograms

Use of Metric Measurement

Metric measurements are generally used throughout, but where Imperial measurements were used in early reports they are also used in these eassays along with approximate metric equivalents.

GLOSSARY

The meanings of words marked with an asterisk (*) are more fully explained in the GLOSSARY before the INDEX.

In Memory

Norman John McPherson 1931-2002

Killed in an aircraft accident at the Lindis Pass, 30 June 2002.
All of the aerial photographs in this book, and in other books in the series, were taken from a plane expertly and co-operatively piloted by Norman during a number of flights over a period of several years.

1.

PATHWAYS TO GOLD

— Roads to the Dunstan

For a newspaper of 1862, the headlines in the *Otago Daily Times* of Saturday 16 August were large and bold:

87 POUNDS WEIGHT
of
GOLD !

Two men, the paper said, had yesterday afternoon dumped 87 pounds weight of gold on the office counter of the startled Dunedin Gold Receiver. They declined to say where they had obtained this rich parcel, but the newspaper speculated that it came from around Mt Watkin near Waikouaiti.

It was not until the Tuesday morning, 19 August, that the newspaper got the story sorted out. The weekend had been taken up by negotiations with the two miners, Horatio Hartley and Christopher Reilly, about the reward that they were claiming for the discovery of a new goldfield. Until this was agreed, they were not disclosing the whereabouts of their find. Finally, with the promise of £2,000, the miners described the location of the discovery as in the great gorge where the Clutha River cut through the Dunstan Mountains.

The newspapers warned would-be gold seekers that it would take about a week to reach the scene of the gold strike on foot, and as it was still early in the season, they could, at the very least, expect bitterly cold nights on the high country, if not snowstorms. The paper emphasised that the interior of Otago was sparsely settled with only a handful of scattered sheep stations. There were no stores of any kind so miners would have to carry all of their supplies with them.

Where was this gorge, soon to be referred to as the Dunstan Gorge, and how did you get there? 'Up the Molyneux[1] River,' the papers said, but where, precisely, practically no one in Dunedin would know. Indeed, only a handful of Dunedin people knew anything of the routes leading into Central Otago.

There were, in fact, three tracks leading into the interior. A northern route up the valley of the Shag River, a central route over the mountains (the 'Mountain Track'), and a southern route through Gabriels Gully[2] and then by way of the rough and indistinct 'Runholders track' up the Clutha River valley.

THE

87 POUNDS WEIGHT

OF

G O L D !

ARRANGEMENT

WITH THE

GOVERNMENT

DISCLOSURE OF THE SPOT!

STATEMENT OF THE DISCOVERERS.

EXTRAORDINARY

RICH FIELD!

DAILY TIMES OFFICE,

MONDAY, AUG. 18, 4.30 P.M.

After some negotiations an arrangement was this day come to between the Provincial Government and the discoverers of the new gold field, and the prospectors made the following statement:—

Figure 1.1 Headlines in the *Otago Daily Times* of 19 August 1862 were unusually bold for the time. This official announcement of Hartley and Reilly's gold discovery initiated the Dunstan Gold Rush.

WAGON TRACKS

With thousands of men starting inland, it was realised by wagoners that food and other supplies would bc urgently needed in large quantities on the new goldfield. On the same day that the first miners from Dunedin arrived at Waikouaiti to begin their tramp up the Shag Valley, Baird Brothers 'New Iron Stores' of Beach Street Waikouaiti, began loading drays with supplies and dispatching them upcountry. These efforts were not motivated by any particularly strong humanitarian desire to help the miners, but because it was judged that the profits to be reaped would be enormous.

The original tracks used in the first few days of the gold rush by foot travellers were not everywhere suitable for the drays that set out to follow the initial crowd. Alternative routes to avoid steep pinches and deep bogs were quickly found, and precipitous stream banks were smoothed down

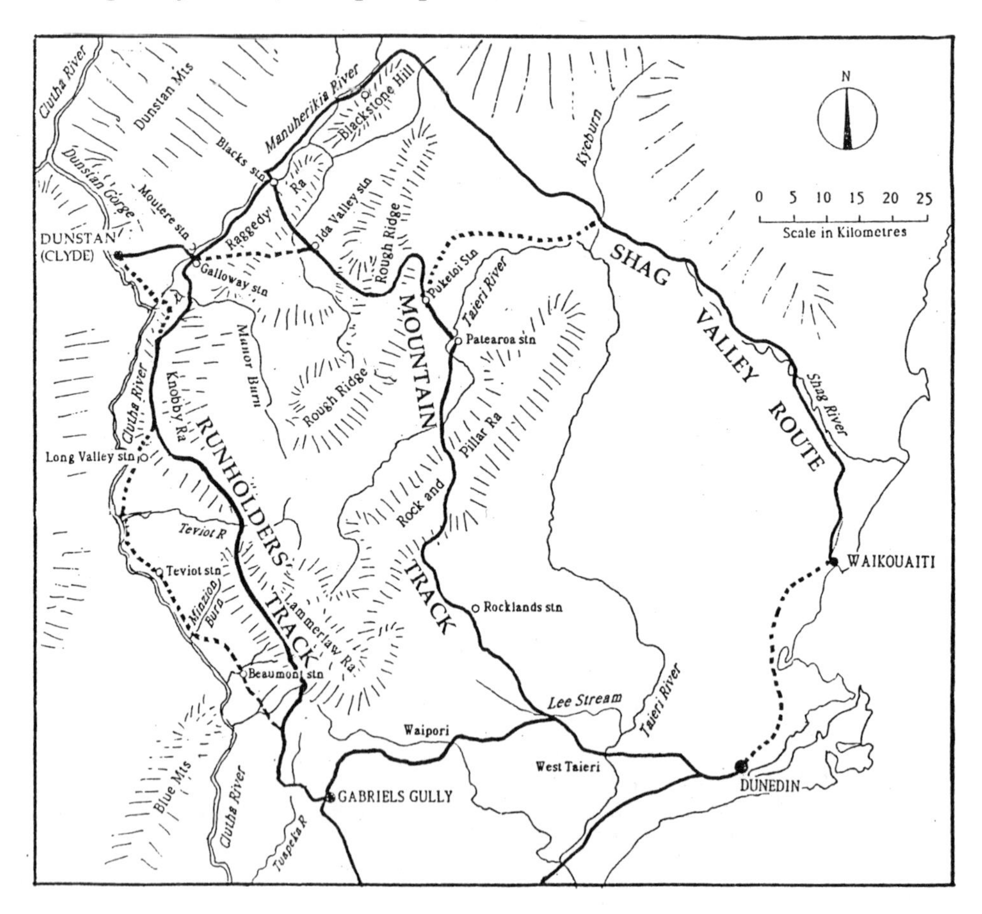

Figure 1. 2. Routes followed by the first drays to reach the new goldfield are shown as solid lines. Routes followed by foot travellers and horsemen were broadly similar, but significant deviations are shown as dotted lines.

by drivers who carried tools just for this purpose.

These heavily loaded drays began to arrive at the diggings just in the nick of time. Some starving miners had already turned back and many others were preparing to force-march back to the nearest food supply. Without this timely arrival of food, the Dunstan Rush may well have petered out. But it takes more than food to sustain a goldfield.

Once the food and drink requirements were under control, the carriers turned their attention to the multitude of necessities demanded by the miners. Soon a constant stream of vehicles, horse-drawn at first but soon supplemented by huge wagons drawn by plodding bullocks, was bringing in tools, timber for mining gear such as cradles and sluice boxes, building materials, furniture and so on.

Figure 1. 3. Two bullock teams prepare to face the steep pull out of Alexandra.

Due credit must be given to the part played by the hard men who wrestled these cumbersome vehicles over the mountains and through the bogs and swampy stream crossings. Sleeping under the wagons in clothes soaked from finding a way through wet tussocks could not have been pleasant in the bitter cold of a Central Otago early spring. Neither was eating rough food which may or may not have been cooked, nor the exhausting physical work of unloading bogged vehicles, lifting and packing the wheels, reloading, the shoving and pushing, swearing and cursing at tired and jibbing horses or straining bullocks. Yes, without doubt they earned the large profits they made.

The routes the vehicles followed can be reconstructed in some detail from accounts left by those who travelled over them, from old maps, and from traces still discernible on the ground and on air photos.

Shag Valley Route

The Murison brothers had formed a track up the Shag Valley as they moved inland in their search for grazing land in 1858. Drays supplying the inland sheep stations used it, and the Shennan brothers of Moutere Station used it to bring their wool out on sledges pulled by bullock teams. Now it was being followed by hundreds of miners. The route began on the coast at Waikouaiti which could be reached from Dunedin by a very difficult track over the bush-clad hills, or more easily, by one of the small vessels which quickly began an 'on demand' service from the town. After a journey of a few hours, passengers were off-loaded into boats manned by local Maori who took them as far as the surf. Here the miners were hoisted on to the backs of husky Maori women who carried them ashore for another fee.

After collecting supplies at Johnny Jones,' or the Baird Brothers' stores, the hopeful miners set off up the Shag Valley,[3] crossed the Pigroot Saddle into the Maniototo Plains and then headed across the plains to Murisons' Puketoi Station. From here they followed Shennans' sledge track over Rough Ridge to Ida Valley Station. Shennans' track climbed the scarp of the Raggedy Range and then ran down a long easy ridge to Low's Galloway Station near the Manuherikia River ford. Once across this river it was an easy 8-mile (12 km) stretch over the high terraces to the entrance to the Dunstan Gorge.

The drays from Waikouaiti worked their way up the Shag River valley in pursuit of the miners, and crossed over the Pigroot Saddle into the Maniototo Plains. But here the routes had to part. Instead of following Shennans' sledge track across the swamps and difficult stream crossings to Puketoi Station, the draymen headed northwest, crossing the northern end of Rough Ridge and Ida Valley. They went around the end of Blackstone Hill, down the Manuherikia Valley to Shennans' station and then across the terraces to the entrance to the Dunstan Gorge. The present State Highway 85 follows much the same route.

On 30 August, five days after the first miners arrived at the Dunstan Gorge, the first wheeled vehicle arrived at the diggings from Waikouaiti. It had left the coast almost at the same time as the miners, and it had taken 10 days for Bairds' dray, with its load of desperately needed food, to get through.

Mountain Track Route

In September 1860 the Provincial Council had voted £2,000 for a road

Figure 1. 4. The steep Brothers hill, with its sticky clay, was a nightmare for early wagoners.
Upper: Wagons, on their way into Central Otago, bogged on Brothers hill.
Lower: The same scene today. Traces of the old road seen in the upper photograph,are still visible near the small patch of scrub at right centre.

from 'West Taieri to the interior.' It was formed during the summer of 1860-61, and was sufficiently complete in April 1861 to allow five drays loaded with stores to pass over it without coming to grief. The 'road' began at Outram, about 17 miles (27 km) across the Taieri Plains from Dunedin, and crossed the plateau to Campbell Thompson's station at Deep Stream (now Rocklands Station). From here it climbed over the high Rock and Pillar Range and then dropped down the steep northwestern face of the range to Valpy's station near Patearoa. This part, known as the 'Mountain Track,' could be a hazardous venture, because most of the 25-mile (40 km) section was over very exposed uplands which reached an altitude of 1,100 metres. A welcome easy stretch of 5 miles (8 km) across the plains brought the travellers to Murisons' Puketoi Station. Here, joined by the foot-sloggers and horsemen who had come up the Shag valley, they followed Shennans' track over Rough Ridge to Ida Valley Station.

Figure 1. 5. The Mountain Track. The climb out of Deep Stream up on to the Rock and Pillar Range was long and steep. The present motor road has developed from the old wagon track.

The fact that the Mountain Track was 30 miles (50 km) shorter than the Shag Valley route, and avoided the double handling required by the sea journey to Waikouaiti, made it attractive to draymen and wagoners.

Travelling in convoy to help each other through the difficulties, the first drays across the Mountain Track arrived at the diggings on 4 September with the news that many more were on their way. By and large the drays had followed the miners along the rough dray track to Ida Valley Station, but there they parted company. The drays had to turn northwards to follow the dray track across the Raggedy Range to Black's Station on the Manuherikia River. Then they turned down the eastern side of the river, picking their way between the rock outcrops until they reached Galloway Station, where they forded the river to Moutere Station and continued on their way to 'the Dunstan,' as the camp at the mouth of the gorge was becoming known.

It is recorded that the Mountain Track carried nearly all of the wheeled traffic during that first summer of 1862-63. But the long hills, particularly the pull out of Deep Stream and the steep climb to the top of Rough Ridge, together with the long tracts of soft boggy soils on the summit of the Rock and Pillar Range, proved to be very hard on horses. It is said that many died on the road. It was quickly realised, too, that the route would be impassable in the winter when as much as three feet of snow lay on it for long periods.

Foot travellers were warned by the newspapers of the dangers of this route because of the initial lack of shelter from bitter winds and snow storms on the long, upland stretches. So mainly horse riders used it, and it remained the preferred summer route for drays and wagons.

Clutha Valley Route

The third route was by way of Gabriels Gully, on the already-established Tuapeka Goldfield. There were at least two ways of reaching Gabriels. The most popular was via the main south road from Dunedin to Tokomairiro River. A reasonable road was then followed inland to the Woolshed diggings at Glenore and on through Waitahuna to Gabriels. The distance was 60 miles (100 km).

The other route to Gabriels headed over the shoulder of the Maungatua Range from near Lee Stream, crossed the Waipori River on a wire bridge and ended at Wetherstons near Gabriels Gully. The track crossed bleak and exposed uplands and was more favoured by wagons than foot travellers.

Across the Lammerlaw and Knobby Ranges

The merchants and carters of the Tuapeka Goldfield were hard hit by the rush to the Dunstan which had seen Gabriels Gully and other nearby diggings practically deserted. Although the runholders' track up the Clutha Valley used by the first rush of miners was impossible for wheeled vehicles, the draymen were determined to find a way through. Several drays had set off to find a new route, and Sub-Inspector Morton of the Tuapeka police was asked to follow them and report on their progress. His report is the best description of the route we have.[4]

After crossing the Tuapeka River, Morton wrote, the drays had climbed the steep spur opposite Bowler's station and continued for many miles up

Figure 1. 6. The Rock and Pillar Range was the most difficult part of the Dunstan. Road.
Upper: A motor road broadly follows the route of the old wagon road across the bleak summit of the Rock and Pillar Range.
Lower: The steep zigzag down the north-western slope of the Rock and Pillar Range follows the original Dunstan Road.

a long leading ridge. A short branch from this took Morton down to a reasonable crossing of the Beaumont River and then he was able to continue along a leading ridge to Beaumont Station. From here he followed the miners' track but instead of turning down the Minzion Burn towards the Clutha River, he turned up the ridge until the dray tracks he was following crossed the stream by a good ford. The tracks led him eventually to the banks of the Teviot River 'near a plain,' almost certainly near the spot where the dam now impounds Lake Onslow. The gradient of the river here was much less than at the dangerous crossing near its junction with the Clutha River, so the current was not nearly so swift and the river was wider and shallower. The only problem was that the approaches on both sides were swampy.

Figure 1. 7. The snow-covered Lammerlaw Range from Mt Teviot. Bullock wagons from Tuapeka found their way across these bleak upland ridges on their way to the Dunstan.

Once across the river, the track headed up the eastern slopes to the summit ridge of the Knobby Range, which it reached near the Pinelheugh. From here it was fairly easy going along the broad summit ridge, although it was high and very exposed. The soil was soft and peaty, but solid rock was only a foot or so below the surface. The only places that gave difficulties were two small streams that had to be crossed and the large swamp at the head of Shanty Creek.

Just to the east of Cairnhill, the foot and horseman traffic branched off to head directly for the Clutha-Manuherikia junction, but the Inspector turned north-east following the runholders' track, and slowly descended the long ridge east of Carters Gully until he reached the Manor Burn. He reported that this crossing, which was some distance upstream from the present lower dam, was very difficult for drays as it involved a very steep descent and an equally steep climb on the other side of the creek. The track then continued along the low ridges and finally came out on Galloway Flat near Low's station. Here it joined the track from Blacks in crossing the river and heading for the Dunstan.

By mid-September, the first dray pulled by two horses and with 16 hundredweights (800 kg) of flour on board, arrived at the diggings from

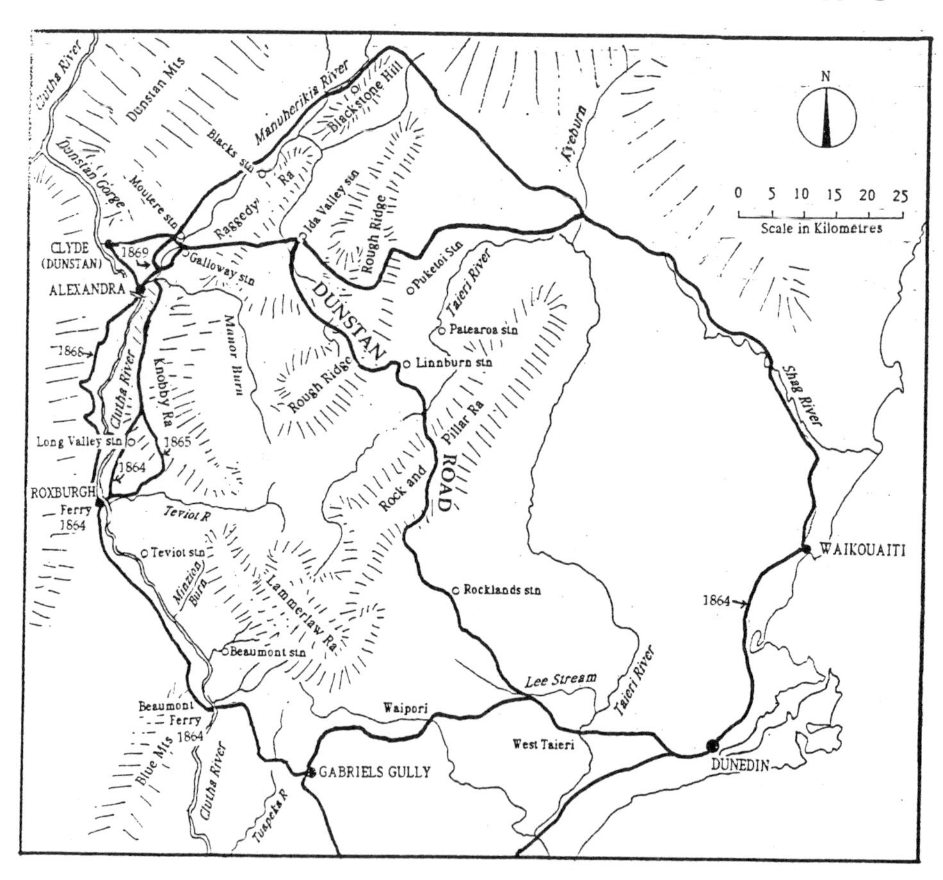

Figure 1. 8. Roads leading into the interior after the period of frantic road construction of the 1860s. The tracks over the Rock and Pillar Range and Knobby Range were 'opened up' to become the Dunstan Road and Knobby Range Road. After the low-level road up the Clutha Valley to Alexandra was opened in 1868, the route over the Knobby Range was abandoned and most of the Dunstan Road phased out.

Tuapeka by way of this route. It had received a considerable amount of assistance from passing miners and their horses.

Other wagoners tried branching off from the newly completed Lee Stream to Waipori road and heading directly up the ridge to the summit of the Lammerlaw Range. Here they had to negotiate the long summit ridge at an elevation of nearly 1,200 metres before joining, near Mt Teviot, the track coming up from Tuapeka.

Although the arrival of drays provided much needed relief for the miners at the diggings, they carried too little to supply the growing demands, not only for food but also for timber for building and for mining equipment. The answer lay in the heavy wagons pulled by large teams of horses, or the slow but powerful teams of bullocks which could bring loads of five tons or more through swamps and up slopes that horses would find impossible.

Captain Jackson Barry tells[5] how he bought five tons of flour at Tuapeka and sent it on its way to the Dunstan. He records that it took the two bullock-drawn drays a fortnight to reach their destination. These high-level routes from Tuapeka proved so difficult that few vehicles attempted them. Not even bullocks could drag huge wagons across the deep trenches in which some of the streams flowed. Nor could top-heavy wagons sidle round steep slopes in safety. Loads were severely restricted and this was reflected in the freight charges, which forced the price of food and other essentials on to outrageous levels. Something had to be done, and quickly.

ROAD BUILDING FRENZY

Fortunately the Provincial Government saw the provision of more easily negotiable roads, and particularly a short, low-level route, as a matter of urgency. The adoption of this policy was no doubt made easier by the large amounts of revenue flowing into the Government's coffers from the tax on the huge quantities of gold being produced.

As early as the beginning of October 1862 a gang of 40 men was sent out to 'open up' [6] the Mountain Track and make it more suitable for drays. Over the summer the Track was largely reconstructed. Some of the improvements included a new ford at Deep Stream, deviations to avoid peat bogs on the summit of the Rock and Pillar Range, and a negotiable zigzag road down the steep north-western face of the range to Styx. From here an entirely new route to the Ida Valley was laid out. It went directly across to Linnburn Station, bypassing Valpy's Patearoa Station and Murisons' Puketoi Station, and so shortened the route considerably.

From Linnburn Station a new road was formed across Rough Ridge south of Shennans' Track, which it joined at Ida Valley Station and followed across Ida Valley to the foot of the steep face of the Raggedy Range. Following Shennans' sledge track up a convenient gully, the new road climbed the range and then took a direct line down a leading ridge

Figure 1. 9. Traces of the Dunstan Road on the north-western face of Rough Ridge.

on to the wide terraces that flank the eastern side of Galloway Flat. Although this section entailed extensive earthworks on the scarp of the Raggedy Range, it avoided the necessity for wagons having to go by way of Black's station. All in all a sum of over £10,000 was spent over 18 months in reconstruction of the Track which was now being referred to as the 'Dunstan Road.'

At the same time, a gang of more than 40 men was working on the Shag Valley route which was much improved by several miles of sideling cuttings. But the main effort on this track was put into the section that crossed the Maniototo Plain to Puketoi Station, where the very swampy crossing of the Gimmerburn was made suitable for wheeled traffic.

Although the Shag Valley road attracted little wheeled traffic during the summer, the work was done in anticipation of the road becoming the main route as winter snows blocked the high-level Dunstan Road. But the route had its own problems. At several places, such as at the Brothers hill and the Houndburn hill, deep mud and slippery conditions made the road impassable in wet winter weather.

A MAIN ROAD

When it was announced that a decision was being made as to which of the routes would become the main road to the goldfields, newspaper correspondents had a field day extolling the advantages and disadvantages of the various roads.

No one seriously thought of the Dunstan Road as a contender because of its elevation and consequent winter snow problems. But the Shag Valley route and that by way of Lawrence were fairly evenly balanced.

By January 1863 the decision had been made—the main road to the Dunstan would run from Lawrence to Beaumont where a punt would carry traffic across the Clutha River. Then the road would follow the western side of the Clutha River up to Teviot (Roxburgh). Eventually it would continue on the western side of the river to Alexandra, but until this difficult section of the road was completed, traffic would cross on a punt at Teviot. After the punt crossing, the road would climb up to the summit ridge of the Knobby Range and follow the existing track to Galloway and so on to the Dunstan.

Although the road from Lawrence would be more difficult to build than the one through the Shag Valley, it would, when completed, lie at a lower elevation and pass through slightly more agricultural land.

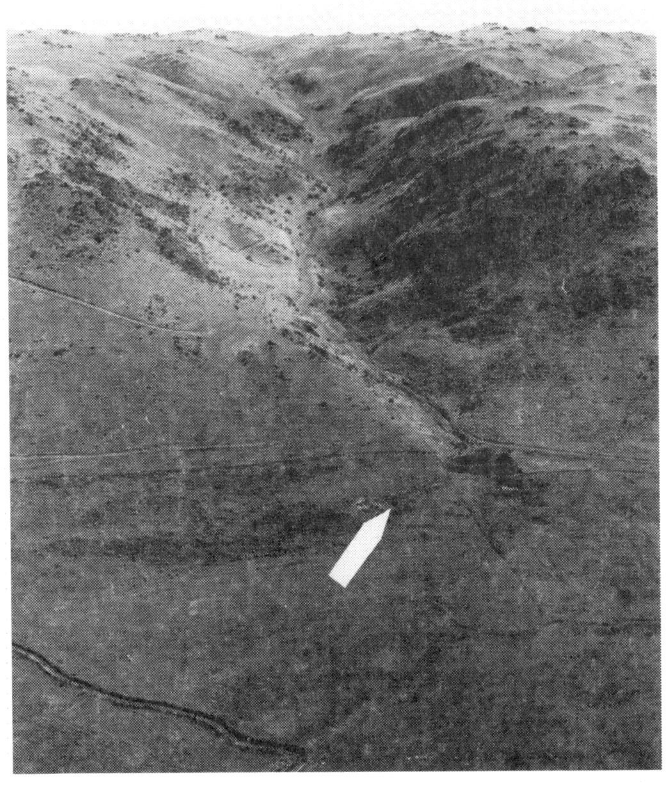

Figure 1. 10. The Dunstan Road climbs the steep south-eastern face of Raggedy Range by way of a gully. The ruins of Baird's Hotel, the 'bottom of the hill' hostelry are arrowed.

Figure 1.11. The road.over the Knobby Range from Teviot is still clearly discernible along the summit ridge.

A gang of 75 men began at Lawrence, and another of 48 at the Dunstan end in an endeavour to push the road through as quickly as possible. Unfortunately the punts took much longer to procure than expected, so the three sections of the completed road from Lawrence to Beaumont, from Beaumont to Teviot and from Teviot across the Knobby Range to Galloway, remained isolated and of little use until the punts were finally installed.

Until the punts were in place at Teviot and Beaumont, wagons and drays still had to use the upland route from Lawrence and a line of cairns and poles was set out to mark the track. But by the end 1864 the punts were operating, and wheeled traffic thankfully abandoned this very difficult route.

After traffic had crossed to the eastern side of the river at Teviot, it headed up the long ridge towards the summit of the Knobby Range. But there were difficulties with this new section of the road. The very steep pinch up on to summit ridge gave wagoners trouble, and a route nearer the Teviot River was brought into use. Snow storms were frequent during winter and the road along the exposed summit ridge was often closed by snow. Nevertheless Cobb and Co. began a coach service in April 1865, but it is said that only one driver, James Carmichael, would take a coach across the Knobby Range road—and when he left, the service was abandoned.[7]

WESTERN APPROACHES

Pressure mounted to extend the road to Alexandra along the western side of the river as soon as possible. Money was allocated in September 1866 but there were delays in starting the work. Nevertheless, it is recorded[8] that a John Fitzgerald was able to drive the first team and wagon directly through to Manuherikia (Alexandra) in June 1868 although the road was not completed until November. Mackersey and Duley had anticipated this by having a very large punt built and in place on the Clutha River at Alexandra before the road was completed.

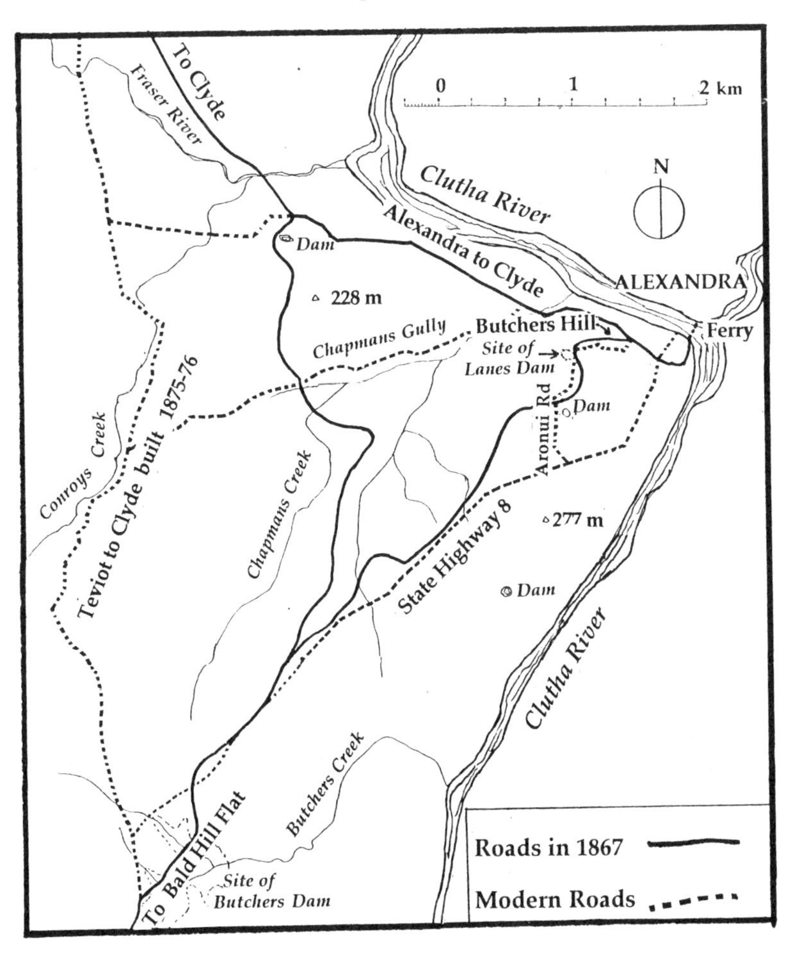

Figure 1. 12. Roads near Alexandra when the road from Roxburgh was opened in 1868 along the western side of the Clutha River.

This new stretch was not easy for the wagoners. A steep pull out of Coal Creek on a sticky clay-based track ('Gluepot hill' it was called), was followed by the tortuous negotiation of deep stream valleys. But these difficulties were nothing compared with the approach to Alexandra. The road picked its way between rock outcrops across what is now Bridge Hill where a road branched off and crossed Chapmans Gully, forded the Fraser River and headed across the stony Earnscleugh Flat for Clyde. The road to Alexandra dropped down a precipitous slope to the present Earnscleugh Road. Butchers Hill, as it was called, was a nightmare for drivers of wagons both going up and coming down. Public meetings were held,[9] and finally tenders called for a new alignment, but even while discussions were going on, John Butler's gig capsized on the hill and his leg was broken.

During 1874 a new road was built southwards from the ferry approach road to Halfmile Gully.[10] It then sidled up the gully and finally climbed out to rejoin the old road—a route still followed by State Highway 8.

Figure 1. 13. This private driveway, off Earnscleugh Road, is almost certainly a remnant of the old road from Roxburgh, opened in 1868, which came down into Alexandra by way of 'Butchers Hill.' Photograph believed to have been taken in the 1920s.

The completion of the road to Alexandra saw the Knobby Range route abandoned, and traffic over the Rock and Pillar Range greatly reduced. Only the stretch of the Dunstan Road over the Raggedy Range from Galloway was maintained as the main access to Ida Valley until the present road over the Crawford Hills was opened.

GALLOWAY INTERCHANGE

As a result of this major road-building effort, three reasonable roads serviced the Dunstan by the end of 1864. They all began at Dunedin on the coast, but followed widely divergent routes through the interior before coming together again in the Galloway-Springvale area a few kilometres up the Manuherikia River from Alexandra.

With the establishment of Manuherikia (Alexandra) as an important centre, roads branched to it from each of the trunk roads, and this led to a pattern of roads around Galloway, which rivalled in complexity a modern motorway interchange.

Most wagon traffic arrived in the district by way of the Mountain Track or the 'Dunstan Road,' as it was becoming known. From the top of the Raggedy Range the road followed down a long easy slope but eventually ran out on to an extensive high terrace. Just before the main road dropped down to the Manuherikia River flats from this terrace, a branch turned south, traversed the length of the extensive Galloway Flat and crossed the Manuherikia River two hundred metres downstream from the present bridge to Galloway. This crossing became known as 'Duncan Robertson's ford.'[11] From this ford the road followed along the foot of the terraces to Alexandra, much as does the modern highway. The main Dunstan Road continued on to ford the Manuherikia River at Moutere Station (in those days, off the present Keddell Road) and then went on to Clyde.

Those wagons that had come up the Shag Valley, across the head of the Ida Valley and down the length of the Manuherikia Valley, joined the Dunstan Road at Moutere Station. But wagons following this route with Alexandra as their destination had to ford the Manuherikia River at Moutere Station, and join the road along the Galloway Flat, recrossing the river at Duncan Robertson's ford.

Wagons, and for a short time the coach, that came over the Knobby Range road from Teviot, came out on to Galloway Flat about midway between the Manor Burn and Dip Creek. Traffic for Clyde continued on past Galloway Station to the Moutere ford, whereas that for Alexandra headed for Duncan Robertson's ford.

It was not long before a major deviation was made to the lower part of the road from Teviot to eliminate the very difficult crossing of the Manor Burn. Instead of following the leading ridge from near Cairnhill down the eastern side of Carters Gully, the new route found its way down ridges to the point where the Manor Burn leaves the hills to flow across Galloway

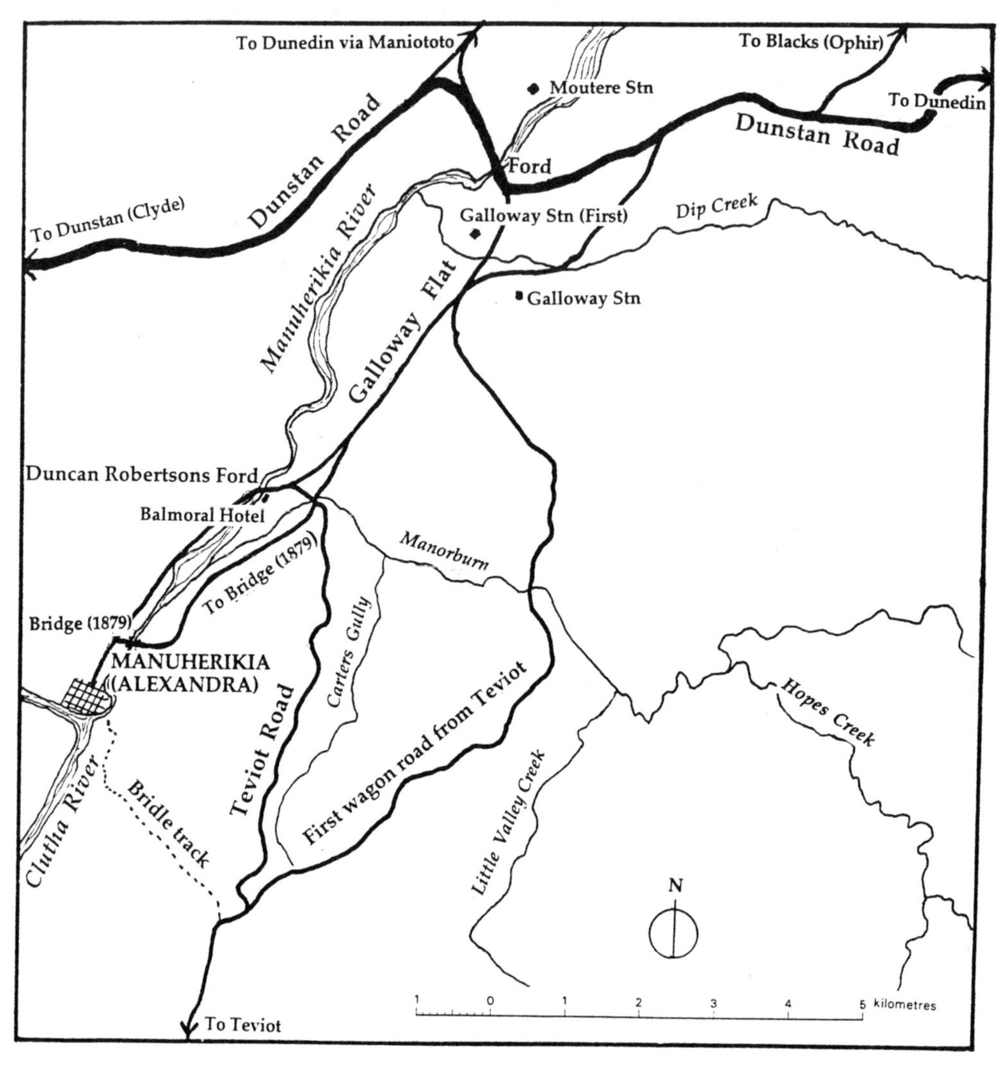

Figure 1. 14. At Galloway Flat, a few kilometres up the Manuherikia River from Alexandra, the three main routes from the coast came together. The pattern was complicated by the need for branch roads from each main road to serve the growing town of Alexandra.

Flat. The new road crossed the Manor Burn here by an easy ford at about the same place as the present Tucker Hill Road fords the stream. Soon afterwards the road branched, with one road running up the flat to the upper ford at Galloway Station and the other heading across to Duncan Robertson's ford. This new deviation[12] quickly became established as the main road to and from Teviot, although foot travellers still used the bridle track commencing at the mouth of Graveyard Gully.

In 1869 a road was completed from Moutere Station across the high terraces to link with the road to Alexandra from Robertson's ford. This meant that wagons coming down the Manuherikia Valley road and bound for Alexandra, no longer had to make the deviation through Galloway Flat which included the double crossing of the Manuherikia River. But many still preferred the river crossings and the easy route along the Galloway Flat to the hills of the new road link.

When the light traffic bridge across the Manuherikia River near Alexandra was opened in 1878, Tucker Hill Road was formed along the terraces on the eastern side of the Manuherikia River to the Manor Burn ford and Galloway. It rapidly became the main route to Galloway and Ida Valley for pedestrian and light horse-drawn traffic as it obviated crossing the river at Robertson's ford.

Galloway, as a hub for the roading system, began to fade in importance only after the opening of the bridge across the Clutha River at Alexandra in 1882, but the network of tracks persisted for many years. In the 1920s it became necessary to lay out roads on Galloway Flat to service farms resulting from cutting up part of Galloway Station. At a meeting of the Vincent County Council the engineer was heard to remark that there were so many roads criss-crossing Galloway Flat that he didn't know where to start with a survey.

Figure 1.15. The road from Teviot near the ford of the Manor Burn. The wheels of heavy wagons have worn ruts in the schist rock.

NOTES

1. Now the Clutha River.
2. Gabriels Gully is close to the present town of Lawrence. Gold was discovered here in July 1861 by Gabriel Read and led to the establishment of the Tuapeka Goldfield — the first successful goldfield in Otago.
3. This route is described by G. M. Hassing, 1911: pp. 35-36.
4. *Otago Daily Times*, 24 September 1862. Morton started on his trip on 10 September and wrote his report on 16 September, so he covered the distance from Tuapeka to the diggings and back in seven days.
5. Barry, J. 1903 p. 71.
6. 'Opening up' a road simply meant making cuttings down to and out of stream beds, cutting narrow sidelings around slopes, making culverts and draining or filling in the worst of the swampy places. Along the summit of the Knobby Range, little more than the placing of poles and cairns to mark the route in snow or fog, was required. The gang of 48 men 'opened up' 18 kilometres of the Knobby Range road in one month.
7. Webster, A. H. H. 1948 p. 81.
8. *Tuapeka Times* 13 June 1868.
9. *Tuapeka Times* 24 July 1873.
10. *Tuapeka Times* 4 February and 13 May 1874.
11. Duncan Robertson owned the Balmoral Hotel that was erected in 1864 close to the crossing. It was destroyed in the 1878 flood. Don, A. 1936 p. 147.
12. With the opening up of Little Valley to settlement after the Great War, part of this deviation was used as a major section of the access road to Little Valley. In many places it was very steep and unsuitable for the growing motor traffic. After a great deal of agitation a new road was constructed up the floor of Graveyard Gully in 1925. This took the place of the steep section leading down to Galloway Flat. This Graveyard Gully road was destroyed by a flood in 1948, and the old road had to be brought into use once more until a new road was formed up the steep face of the Knobby Range overlooking Alexandra. This new road joined the old a couple of kilometres beyond Observation Hill.

2.

GOLD IN THE SNOW

—Mining at Campbell Creek

It was the search for the mythical 'mother lode' that drove prospectors right up to the top of the Old Man Range. By 'mother lode' they had in mind a huge mass of quartz* so rich that it was practically pure gold. They believed that the large quantities of gold in the gullies of Conroys, Butchers and Omeo Creeks could only have come from such a source. When the more adventurous among them crossed the range, which was the boundary of the Dunstan Goldfield, and began to prospect gullies on the western side, the newspapers had difficulty in following their activities. This was exacerbated by the miners' reluctance to disclose information, because their prospecting was by then outside the legal boundaries of the Goldfield. This meant they could not obtain the protection of a licensed claim.

CAMPBELLS GULLY

There were various newspaper references during November 1862 to payable gold having been found 'behind the Snowy Ranges' (the Old Man Range), but it was late in the afternoon of 6 December 1862 when:

> . . . A. Campbell[1] and party applied for a prospecting claim in a gully fully as large and almost resembling Gabriel's. Just as they had bottomed and succeeded in striking gold, they were discovered, and had to rush in immediately to secure an extended claim. Prospect obtained, 6 dwt [= 6 pennyweight = 6/20 of an ounce]. The men believe they have found ground that will employ eight to ten thousand people. It is eighteen miles from the camp [Dunstan] in a southerly direction and in a portion of the country very little tried.[2]

Apparently the early arrivals were able to deceive others about their returns for some time, but eventually in May 1863,[3] the place was rushed and the rush continued for some weeks. A crowd even left Alexandra late in July for the diggings in spite of the fact that heavy winter snows had already fallen.

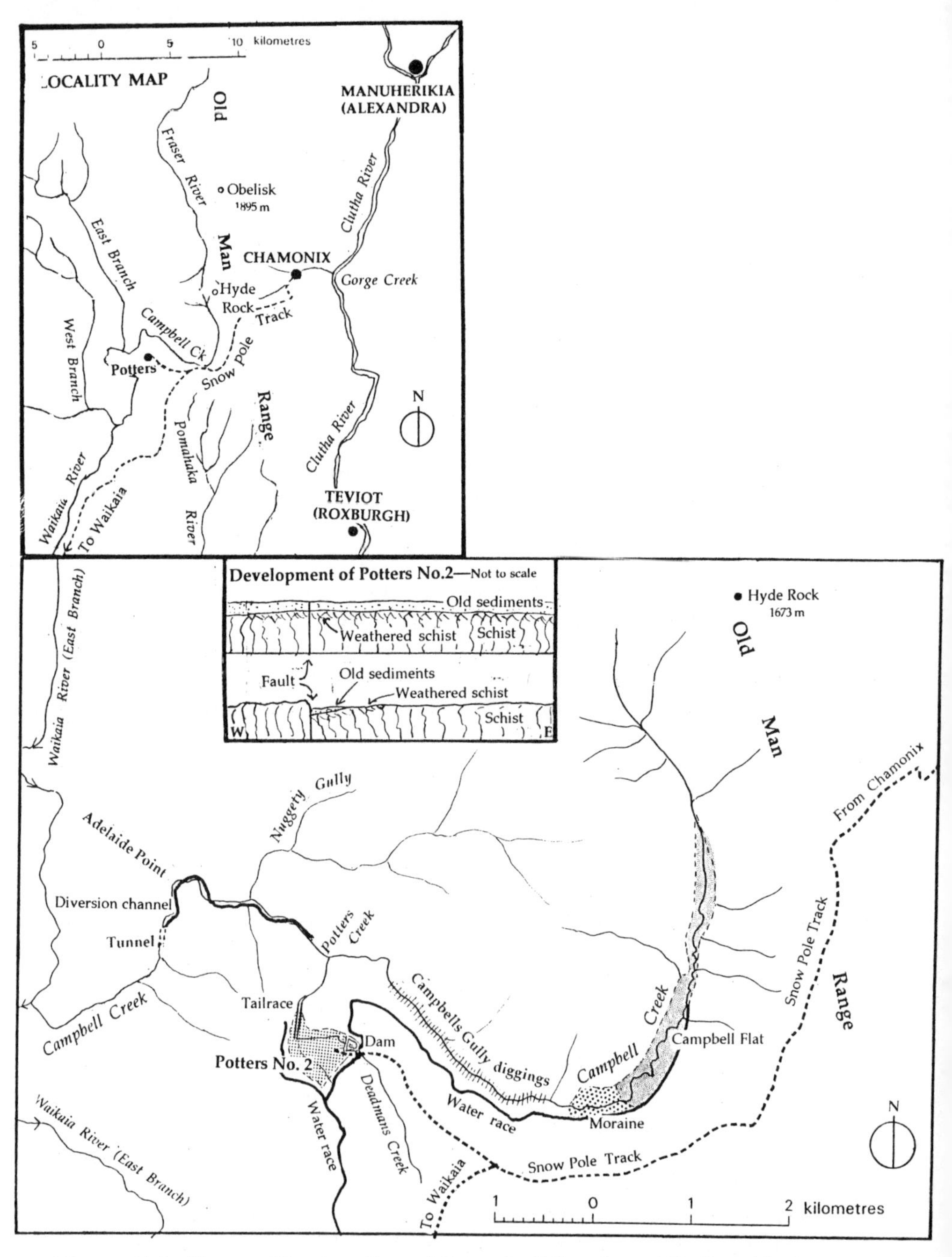

Figure 2.1. Upper: The location of Campbell Creek and Potters No 2.
Lower: Diggings around Campbell Creek
Inset: Potters Gully lies in a small down-faulted basin.

already fallen.

Campbell Creek rises close to Hyde Rock, a prominent tor* on the summit ridge of the Old Man Range, and flows down the western slopes of the range into the Waikaia River. The almost straight course and smoothed-off walls of the upper valley indicate that, in past time, this part was modified by glacial action. Further evidence of this activity is a small terminal moraine* about four kilometres below the source. As the small glacier, which occupied the upper valley, melted, the moraine obstructed the drainage and a long narrow lake formed behind it. This has since been completely filled in to form a nearly level alluvial plain, across which the present stream meanders. It was called 'Campbell's Flat' by the miners. There is no record of gold being found on this flat, nor in the moraine, presumably because of the great thickness of overlying debris.

Figure 2. 2. Upper Campbell Creek valley has been glaciated. A small terminal moraine obstructed the drainage and formed a long narrow lake which has silted up to form Campbell Flat (beyond the moraine).

However, for about two kilometres between the moraine and the entrance to the gorge through which the stream flows on to the Waikaia River, the narrow valley of Campbell Creek was the scene of intensive gold diggings. It was here that canvas stores and miner's tents made up the township of 'Campbells Gully.'

Although the diggings at Campbells Gully yielded good returns, the area of ground available was limited, and newcomers were soon prospecting the surrounding country. Almost immediately gold was discovered about two kilometres to the southwest in a shallow, peaty basin near the head of a small stream draining into Campbell Creek. It was called 'Potters Creek' after the discoverer.

Figure 2. 3. Campbells diggings are between the moraine (out of picture to right) and the entrance to the gorge (background).

POTTERS No 2

The diggings in Potters Creek were called 'Potters Gully No. 2' because it was the second golden gully John Lishman Potter had discovered.[4]

Potters No. 2 is a broad shallow basin covered with peat, similar to scores of peaty basins in the vicinity, but different in one important respect — on its west side the basin is bounded by a long, slightly curved scarp about 40 metres high. This scarp indicates that the basin has been formed by the land dropping down along a fault.* The lowering of the land in this way has protected from erosion a remnant of the ancient quartz gravels and sands that once covered much of Central Otago.

Coarse gold, including small nuggets, was found at shallow depths beneath a layer of schist* gravel that had been washed into the basin from the nearby slopes. In the northern part of the basin the gold lay on soft weathered schist rock, but in the southern part it lay on the old quartz sands which themselves contained some gold.

Gold was won by cradling* and by ground sluicing.* Water was brought from Campbell Creek[5] in a long water race, and another picked up water from a number of small streams (mainly from Three Ounce Creek) lying to the south of the basin. Water from both races was stored in two reservoirs. One was formed by a dam built across Deadmans Gully, which drains into Potters basin, and the other was formed by building a sod wall across a small depression on the western side of the basin.

Figure 2. 4. John L Potter discovered gold at Potters No. 2.

Most mining was done by groups of men working as partners. One party of eight miners from Teviot, headed by J. Chatburn, and calling itself the 'Hit or Miss Company,' spent a year cutting the six mile (10 km)-long water race from Campbell Creek to Potters Gully. Then they spent another two years blasting a tail race,* up to two metres deep, through solid rock for a distance of more than a quarter of a mile along the line of the outlet stream. Water, stored in the dam they had built across Deadmans Gully, allowed them to divide their party into a day and a night shift.

Mining in Potters Gully petered out about 1873, although one or two solitary miners continued working into the new century.

THE GORGE

As the narrow, flat valley bottom in the upper part of Campbell Creek was worked out, miners moved down stream into the deep gorge. Payable gold was found at a number of places including one called 'Adelaide Point' which became an important mining locality. In fact by 1865 Adelaide Point and Potters were said to be the principal mining locations at

Figure 2. 5. Air view westwards across Potters basin to the fault scarp. Tail race runs out to right. Outcrop of old sediments (sands) top left. Diggings centre. Road from SH 8 ends at huts.

Campbells Diggings. Unfortunately, there is doubt about the precise locality of Adelaide Point and Adelaide Gully, a gully presumably somewhere nearby. Their position is not clearly described nor shown on any maps. We are left to deduce their position by piecing together scattered and vague information.

Adelaide Point and Adelaide Gully

The first mention of Adelaide Gully is in September 1863 when a newspaper correspondent visited the diggings, which were under deep snow. From Potters he:

> . . . crossed a hill, and about a mile distant I came to another working, called Adelaide Gully, where there were about thirty diggers. The sinking was shallow, and all were working briskly amidst the snow. . .[6]

In early February 1864 a rush to Campbells Gully 'near the junction of Potters No 2 Gully' was announced, but a report a few days later described the event as:

> . . . a rich find at Adelaide Gully spur situated about the lower end of Campbells Gully.

The most obvious spur in the gorge is nearly a mile (1.5 km) down stream from the junction of Potters Gully where the creek is forced to turn through 180^{0}. It is likely that this was 'Adelaide Point.' Later a number of substantial miners' huts were built on the end of the spur (the 'Tunnel' huts).

A report in the *Tuapeka Times* gives another clue to the locality of Adelaide Point:

> Toy and Savage have just cut a 600 ft tunnel through a slip at Adelaide Point. It has taken them three years.[7]

The approximate position of this tunnel, which was built to act as a tail race* for mining operations in the bed of Campbell Creek, is known, though all signs of it have now been obliterated. It was about 600 yards down stream from 'Adelaide Point.'

The Tunnel Party

No doubt the rich returns from the vicinity of Adelaide Point encouraged a party of miners to begin this ambitious project which was to occupy them for the next 40 years. Their intention was to mine the entire width of the floor of the gorge, including the bed of the creek itself. But first they had to form a tail race with sufficient fall to get rid of their tailings.* They achieved this by driving a tunnel, about 200 metres long, through a slip which had fallen from the side of the gorge and obstructed the stream. It is likely that this tunnel was 'the costly tail-race being cut at Adelaide Point'[7] which was the subject of a co-operative agreement, drawn up in February 1864, by no fewer than 18 miners. They agreed that they would share the labour and expense of cutting the tailrace which they would all

Figure 2. 6. A tail race up to 3 metres deep and half a kilometre long, was blasted Through rock to allow mining of Potters basin in the background.

Figure 2. 7. The Tunnel Party's flood race.
Upper: Excavated In the flood plain
Centre: Excavated in hillside.
Opposite: Built-up on flood plain.
Diagrams on opposite page explain construction

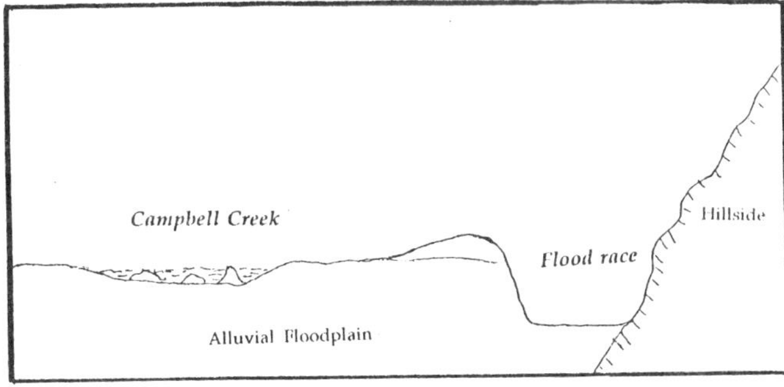
Campbell Creek
Hillside
Flood race
Alluvial Floodplain

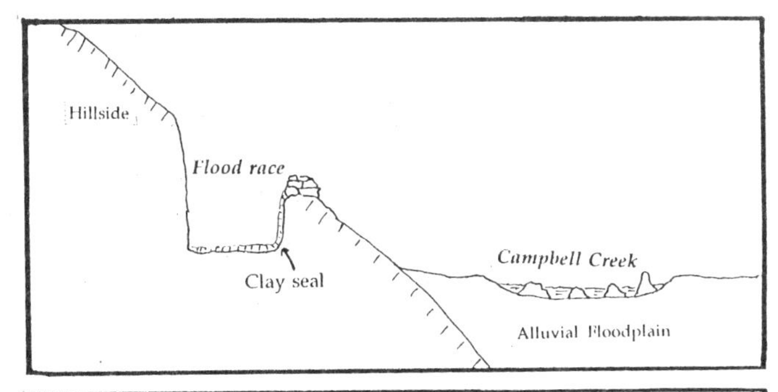
Hillside
Flood race
Clay seal
Campbell Creek
Alluvial Floodplain

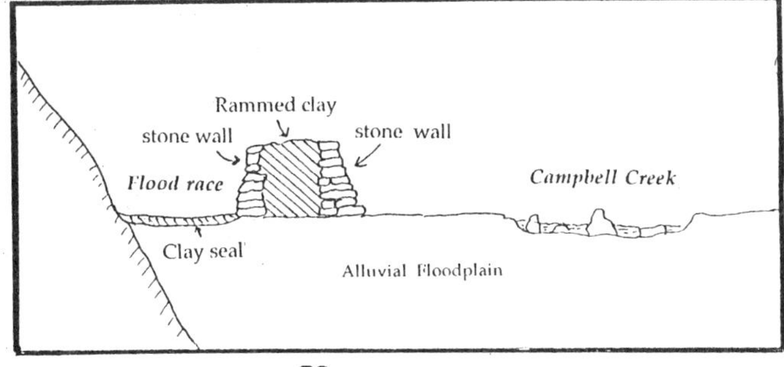
Rammed clay
stone wall
stone wall
Flood race
Campbell Creek
Clay seal
Alluvial Floodplain

use to dispose of their tailings. They also had to agree that they would work every day on the tailrace, unless ill or away on urgent business. If otherwise absent they had to pay £1 for each day they were away.

When their tunnel was finished, the 'Tunnel Party' as they were called, diverted a section of the stream, about 100 yards in length, into a wide channel they had constructed alongside the creek, thus exposing the original stream bed for mining. When they had worked out the section, another length of stream was diverted and the process started again.

It was a work of some magnitude. Where there was sufficient room on the floor of the valley, a ditch about six feet wide and three feet deep was dug to accommodate the stream. But where the valley was narrow and mainly occupied by the stream, the flood race* had to be taken along the adjoining hillside, supported by stone walls where necessary, and through the ends of ridge spurs in deep rock cuttings. Sometimes it was easier to build the flood race above the flood plain rather than excavate a ditch. Then a massive bank was built about six feet out from the hillside by constructing two parallel stone walls and filling the space between with rammed clay. The diverted stream ran between the bank and the hillside.

Over the years the party mined several miles of the gorge and operations were only halted, apparently, by the narrowing of the valley and the presence of huge boulders. During the long life of the claim, the personnel of the Tunnel Party changed, as might be expected, but some who had joined in 1864 were still working 35 years later. The men lived in well-built stone cottages with thatched roofs, high on the hillside at 'Adelaide Point.'

GETTING TO CAMPBELLS

Miners at Campbells diggings had to face very severe difficulties. Firstly, there was the problem of access. Every item of equipment and food had to be carried in, either on the miners' backs or by packhorse. Supplies were scarce in the early days of the discovery, but this was remedied as local packers from Manuherikia and Teviot began to make trips across the range with supplies. After the rushes of mid-1863, the population expanded quickly and a number of stores were opened. Bringing in supplies became big business. It was not long before a bustling tent village grew in the deep valley of Gorge Creek on the eastern side of the Old Man Range about 12 miles (20 km) from Alexandra. It was called Chamonix[9] and if ever there was a single-industry village, this was it. There were about 20 stores, shanties and accommodation tents as well as a blacksmith shop, and even a skittle alley, all jammed together along a narrow steep street.[10] The whole place was devoted to packing stores up to miners at Campbells, and to the diggings on the Fraser and Pomahaka Rivers.

Figure 2. 8. Site of Chamonix. Evidence of tent sites strongly suggests that the packers' village was in Hut Creek (the northern branch of Gorge Creek) on the clear area in the centre of the photograph. The route to Campbell Creek crossed the main branch of Gorge Creek and climbed the steep slope to a leading ridge (out of the photograph to the left) and followed this to the summit of the Old Man Range (on skyline)

When there was snow on the ground at Campbells, there was no overnight feed available for horses so the return journey had to be made that same day. Each packhorse carried 100 to 130 kilograms of useful load as well as a certain amount of corn for its own feed.

There was an hour's stiff climb out of Gorge Creek to the shanty at the Springs located on the steep leading ridge, which the track then followed nearly to the top of the Old Man Range. Some distance above the Springs, the Government-erected line of snow poles began. These poles, 45 in all and spaced 200 yards (180 m) apart, were each about 16 feet (5 m) high and painted dark red, with a cross-board near the top pointing to the next pole. They were erected during the autumn of 1863 when it was realised that even a minor snowfall could make direction finding very difficult.

As tussock gave way to alpine herb field, with its carpet of celmisias and gentians, the slope eased off towards the highest point of the track that was at an elevation of 1,590 metres. It was here that it was believed the snow poles would be of greatest service in guiding travellers across the bleak, featureless summit plateau. The gentle downhill slope encountered on the western side of the range would have been more welcome if it had not harboured extensive peat bogs into which the miners sank up to their shins.

LIFE AT CAMPBELLS

The other great difficulty that the miners at Campbell's diggings had to contend with was the climate. Not only were the diggings 1,200 metres above sea level, but they were on the exposed side of the first substantial mountain range to lie across the path of squalls sweeping up from the Southern Ocean. Even in summer, sudden storms, accompanied by

Figure 2. 9. The root of the ferocious speargrass *(Aciphylla aurea)* was recommended as an emergency food but had certain limitations.

Antarctic temperatures, sweep across the high ridges leaving a smattering of snow. What appeared to the citizens of distant Alexandra township to be fog covering the summit was, more often than not, a blizzard on the mountain.

During the summer, life under canvas was not too bad, although it could be very hot living and working in the narrow gullies, but as the season advanced, nights became chilly and heavy frost became common. In the depths of the gorge, miners were lucky if they saw the sun for two hours a day. It was then that the scarcity of firing became manifest. Certainly there was a beech forest further down the western side of the range, but to collect firewood there meant a journey of many miles. Otherwise there were the thin twigs of 'Wild Irishman' (matagouri[11]), which burnt very quickly, and fires required constant attention. There was plenty of peat on the gently sloping summit ridges, but by the time the miners realised its value for fuel, the season was too far advanced to allow adequate drying of the cut blocks. Some, indeed, were so unfamiliar with the material that they built the fireplaces and chimneys attached to their tents from the easily cut sods, and were greatly surprised when the structures caught fire.

During the mild early winter of 1863 the population grew rapidly, with one rush as late in the season as 28 July, until there were about 150 in Campbell Creek itself, and perhaps another 350 in the surrounding high country. Several stores had been established, including one by James Rivers, well-known storekeeper of Alexandra, as well as those of a baker and that of Charles Nieper, a butcher. But often snow storms prevented the packers bringing in supplies, and it was then that the advice given in the newspaper for cooking speargrass (*Aciphylla aurea*) became useful. It was said that the root, which resembled a parsnip, should be boiled for two minutes, scraped and then baked in the ashes. It was said to be very sweet but there was a warning:

> Too great a use of this edible should not be made at first, as it produces a laxity of the system.[12]

Snow began to fall in July and parties of miners, finding conditions too unpleasant, moved down into the shelter of the heavy bush in the Waikaia Valley or pushed on out to Switzers (Waikaia). Others crossed the range to Chamonix but a large number stayed on. It is said that it was the fear of having their claims, which were not yet protected by law, 'jumped' or having them forfeited to another party that caused miners to remain. There was also the reluctance to abandon claims that were producing good gold.

When packers could not get through regularly, stores began to run short of provisions, and few miners had any food reserves (experienced men always kept a large bag of flour in their tent as a reserve). The climax came on 11 August when very heavy snowstorms began to sweep over the range.

Some idea of the amount of snow that fell at Campbells during the storms in August 1863 is given by the report of the newspaper correspondent who tramped into the diggings only a month later. He thought the snow on the 'great glacier,' varied from eight inches to six feet (20 cm to 2 m) in depth, and when he arrived at Potters No 2 everything was still covered by snow about four feet deep. In spite of the snow, 60 or 70 miners were at work, clearing the snow from their 'paddocks' or digging through it to reach water races, which continued to run under the snow. Water from buried races appeared to emerge from snow-tunnels into the sluice boxes. Some miners were actually working in caves they had dug under the snow

Conditions were not unpleasant as long as the sun was shining, but it became bitterly cold at night. A graphic description of how a miner prepared for bed on a cold evening at Campbells is left by the newspaper correspondent:

> A miner, when he turns in for the night, puts on his monkey jacket, his comforter round his neck, ties his fur or sheepskin hat over his head, rolls his feet in a pair of blankets, lights his pipe, lays down, and blows his cloud, and smokes himself off into a balmy sleep.

Those who had dug into the hillsides to form a sheltered platform for their tents, suffered as snow drifted and slid down the slope and buried the tents to a depth of 10 feet (6 m) or more. Many tents were crushed and others survived only because the owners worked all night clearing snow as it fell. Little wonder that a number of the men, inexperienced in the ways of the mountains, already short of food, and knowing nothing of the edible qualities of the speargrass, began to panic. Fearing starvation if they were to remain cut off by the falling snow, they decided to break out to Chamonix. But they had left it too late.

Disaster

Experienced people in Clyde and Alexandra guessed what was going on up on the Old Man Range. They could see the dense black 'dropping cloud' hanging 'like a funeral pall' over the summit of the range, and when it lifted for a moment they could see the shining mantle of newly-fallen snow.

Those miners who made it to Chamonix reported that others were following, but when they did not turn up by 14 August, those who knew the mountain were greatly concerned.

These fears were confirmed on Monday 18 August when exhausted men staggered into Chamonix with tidings of disaster. Parties of miners, trying to break out from Campbells had been caught by a blizzard on the 'great glacier' — the miner's term for the broad snowfield on the summit ridge of the Old Man Range. Many were dead. Caught in whiteout conditions with visibility at zero and all sense of direction lost, they had wandered until overcome by freezing temperatures and exhaustion. A few packers had saved themselves by jettisoning their loads and letting their horses lead them to safety.

Relief teams were immediately organised in Alexandra and they scoured the slopes of the mountain, bringing in a number of badly frostbitten and partly delirious men. They were the lucky ones. For those caught on the summit there was no hope.

How Many Died?

No one knows exactly how many were lost in the August storm. Warden Robinson, who was stationed at Clyde at the time, wrote 24 years later that the number was estimated at 30.[13] James Sandison, an experienced miner, was asked by Vincent Pyke the Gold Fields Secretary, to inspect the route to Campbell Creek before the next winter, and make suggestions as to how it could be made safer. In his report[14] Sandison stated that seventeen bodies had been found in the immediate vicinity of the snow-pole track, but it is likely that he included in this number those who had perished in snowstorms throughout the year. He mentioned that another 19 people were listed as 'missing,' but such was the difficulty of trying to identify miners who were often only known by a nickname, that it is possible that most of these had simply left the district.[11]

Figure 2. 10. New poles have recently been placed in some of the original stone cairns to mark the old snow-pole track.

Several attempts have been made to reach a more accurate figure for the number of deaths during the storm. A writer in 1928[15] stated that only 8 lives were lost and six of the bodies were buried at 'Shingle Creek' (he meant Gorge Creek), but the article has so many other errors that it is doubtful if it can be relied on. More recently Wood carried out a thorough search of the records and could find evidence of only five deaths during the storm. He was of the opinion that the particular storm was nothing unusual for the locality, and that the deaths were caused not so much by the severity of the storm, as by the inexperience of the miners.[16]

Aftermath

Modern readers, used to having disasters reported at great length and in gory detail complete with interviews of survivors, followed by analyses of causes and apportioning of blame, would find the coverage given to the deaths on the Old Man Range strangely scanty. The Dunstan correspondent of the *Otago Daily Times*, under headlines that announced "Five Hundred Men Snowed-up," reported on 21 August that heavy snow had fallen on the ranges and 'great fears are entertained for the safety of several parties who ought to have turned up ere this.' Three days later there was a short account of meetings held to dispatch rescue parties, but most of the article was devoted to people snowed up in other places. And that was it.

The comprehensive report on the Goldfields submitted by Vincent Pyke to the Provincial Government in November 1863[17] put the event into perspective:

> It was recently reported that 500 men were completely snowed up at Campbell's but, on enquiry, it was found that the track was only closed for a very short time. Many similar rumours have an equally slight foundation in fact. It is undoubtedly true that human lives were lost in the floods and snowstorms of the season through which we have just passed; but it is no less true that numbers have been grossly exaggerated. It is to be hoped that our miners, warned by the sad experience of their first winter in the interior, will avoid similar mischance in future years by adopting more precaution in their movements. . .

Although it seemed unsympathetic, the report was essentially true. The loss of life, much exaggerated, had been caused, not by the exceptional ferocity of the storm but by the miners' inexperience of Central Otago winters, and the lack of appreciation of the dangers of storms on the bleak uplands.

Criticism was levelled at the snow poles. They should have been higher; painted black; set in a cairn of stones; have some indication on them of the direction in which to proceed, and they should have ended at some place of safety and not just stop in the middle of a wilderness. From the last pole (No 45), there was nothing but bleak peat bogs stretching more than four miles (6 km) to Potters No 2. After his inspection, James

Figure 2. 11. Remains of the stone shelter huts on the Old Man Range.
Upper. Highest point of the snow-pole track.
Lower. On the summit ridge at the head of Shingle Creek.

Sandison recommended in February 1864 that a line of galvanised wire should be installed 4 ft 6ins (1.4 m) above the ground, supported by posts 45 yards (40 m) apart and carried right to Potters No 2, a total distance of 7 miles (12 km). Also two shelter huts should be built, one at the Springs and one 'at the last rocks on the west side of the ranges.' He also pointed out that winter was only six weeks away so it was time for urgent action.

The Provincial Government certainly acted. Between March and May, James Whisker, a contractor, installed 300 poles, each 10 feet (3 m) long, set in a cairn of stones, at £2 per pole. Five miles (8 km) of wire were stretched along these at a height of six feet above the ground, a height which it was hoped would leave the wire above any snow. Whisker also got the contract to build two shelter sheds in stone, one of which was built at the highest point of the snow-pole track and the other on the summit ridge at the head of Shingle Creek.

Figure 2. 12. The memorial at Gorge Creek to those lost in storms on the Old Man Range.

When the first severe snowstorm of the 1864 winter arrived in June, the sheds were already completed and the snow-pole line, with its wire, was finished. The snow was estimated to be nearly six feet deep in the shallowest places and most of the poles were buried. The wire, where it could be seen, was coated with ice 6 inches thick and was sagged down into the snow. When the snow melted, the wire was found broken into small pieces, probably by frost. It was not replaced.

The deaths during the winter snow did not seem to seriously detract from Campbell's popularity as a goldfield, as by 1865 there was still a population of 126, of whom 107 were miners. But for most miners, activity was confined to the eight months between October and May. During the winter months, the miners worked on claims along the Clutha River and elsewhere. Those who chose to stay in the mountains were those who had built stone cottages and made certain they were well stocked with food.

The story of the large number of miners reputed to have perished in the 'great' snow of August 1863 has entered the folk-lore of Central Otago. Never mind whether or not it is entirely true, the message it conveys is still clear — the bleak, high uplands of the Old Man Range, with their sudden and drastic weather changes, are not to be taken lightly. They are just as dangerous to the inexperienced tourists of today as they were to the inexperienced miners of the 1860s. The monument to the lost miners erected in 1928 at Gorge Creek should be a constant reminder of the fact.

NOTES

1. John Hall-Jones (1982) mentions 'Little Johnny' Campbell as the discoverer of the field, and Peter Chandler (1984) says local tradition supports John Campbell or 'Campbell Brothers.' On 24 September 1863 the *Otago Daily Time's* correspondent, reporting on a visit to the diggings, mentioned meeting 'the prospectors (Campbells).' Used in this sense 'prospectors' meant the discoverers of the field.

A published list (*Otago Daily Times* 5 September 1863) of mail awaiting collection at Manuherikia lists John H. Campbell and Alex J. Campbell. Perhaps all the suggestions are correct — Campbells Gully was discovered by, and named after, the Campbell brothers, Alex and John.

2. *Otago Daily Times* 9 December 1862.
3. AJHR D - 6 1863 p. 6.
4. His first discovery, in a gully first named 'Potters' but later 'Potters No 1,' is a tributary of the Nevis River.
5. Miller, F. W. G. 1966 pp. 22-23.
6. *Otago Daily Times* 24 September 1863.
7. *Tuapeka Times* 26 December 1874.
8. Departmental Reports Session XX 1865 Votes & Proceedings p. 82.
9. There are a number of different spellings, eg 'Chamouni' *Otago Daily Times* 23 September 1863; Robinson *in* Pyke: 1887; ' Chamounix,' Gilkison: 1930,

Sumpter: 1947, Webster: 1948, and others; *Chomonix* Salmon: 1963; Wood 1970. Presumably it was named for the French alpine town, Chamonix.

10. Glasson, H. A. 1957. pp. 114-116. Glasson's description is based on an article written by a newspaper correspondent who visited the village and climbed the range to Campbell Creek in September 1863 (*Otago Daily Times* 23-24 September 1863) This article contains almost all the information we have about this village. The exact location of the village was uncertain until an archaeological study revealed ground markings indicating tent sites on a sloping terrace remnant in Hut Creek. the northern branch of Gorge Creek, a short distance above its confluence with the south branch (Stone, 1996 and pers. comm. Ms Jenny Stone of Arrowtown).

11. 'Matagourie' is the settlers' version of the Maori name Toumata-kuru (*Discaria toumatou*).

12. *Otago Daily Times* 21 August 1863.

13. In Pyke 1887 p.101.

14. *Otago Daily Times* 3 March 1864.

15. *Alexandra Herald* 29 August 1928.

16. Wood, S. L. n.d. Extract as Appendix 1 *in* Hamel G. E. 1989. It is true that the snowfall was probably not exceptional, but the winter was probably more severe than those experienced today. Apart from the heavy August snowfall on the ranges, 1863 was a year of successive and disastrous floods which caused great damage and loss of life in the Wakatipu district. 1863 was close to the end of the stormy, cold period described by climatologists as 'the Little Ice Age.'

17. Pyke. V., Votes & Proceedings Otago Provincial Government. Council Paper, Session XVII 1863. Second Annual Report on the Gold-Fields of Otago, New Zealand, p. 17.

3.

THE BATTLE OF BALD HILL FLAT

Miners v Farmers

The main highway to the south climbs steeply out of the Clutha Valley at Alexandra, and then runs for several kilometres across a plateau peppered with upstanding tors of schist rock. After about ten kilometres, the rocky landscape suddenly gives way to lush flat pastures separated by regular, parallel rows of poplar trees. The undulating highway levels out and runs with unusual straightness for four or five kilometres. There are some modern houses to be seen but there are also many old stone buildings, some derelict, others restored and occupied.

Soon the highway runs down to cross Obelisk Creek and the barren, rocky landscape, typical of Central Otago, reappears. We have just passed through the small farming district quaintly known as 'Fruitlands.'

The Name

The district was not always known as 'Fruitlands'. When the gold discovery was made here in 1863 it was said to be at 'the southern end of Twelve-mile Flat.' Presumably this was the distance from the Dunstan (Clyde), the centre of the universe at this time. But soon the name 'Speargrass Flat' was being used for the diggings that were mainly on the flood plain of Obelisk Creek.

John Kemp, an early miner and hotelkeeper, claimed[1] it was he who was responsible for renaming the place 'Bald Hill Flat' in 1865. He gave his address as 'Speargrass Flat' at the Post Office, but was told that there was already a Speargrass Flat near Arrowtown.

"All right then. We'll call my place Bald Hill Flat."

There is no difficulty in identifying the 'bald' hill that Kemp had in mind when he suggested the name. It is the prominent conical hill standing about 25 metres above the plain on the eastern side of the main highway about a third of the way along the Flat from the northern end. The 'bald' appearance was caused by a cap of porous gravel at the top, which dried out and so couldn't support the tussock vegetation of the lower slopes.

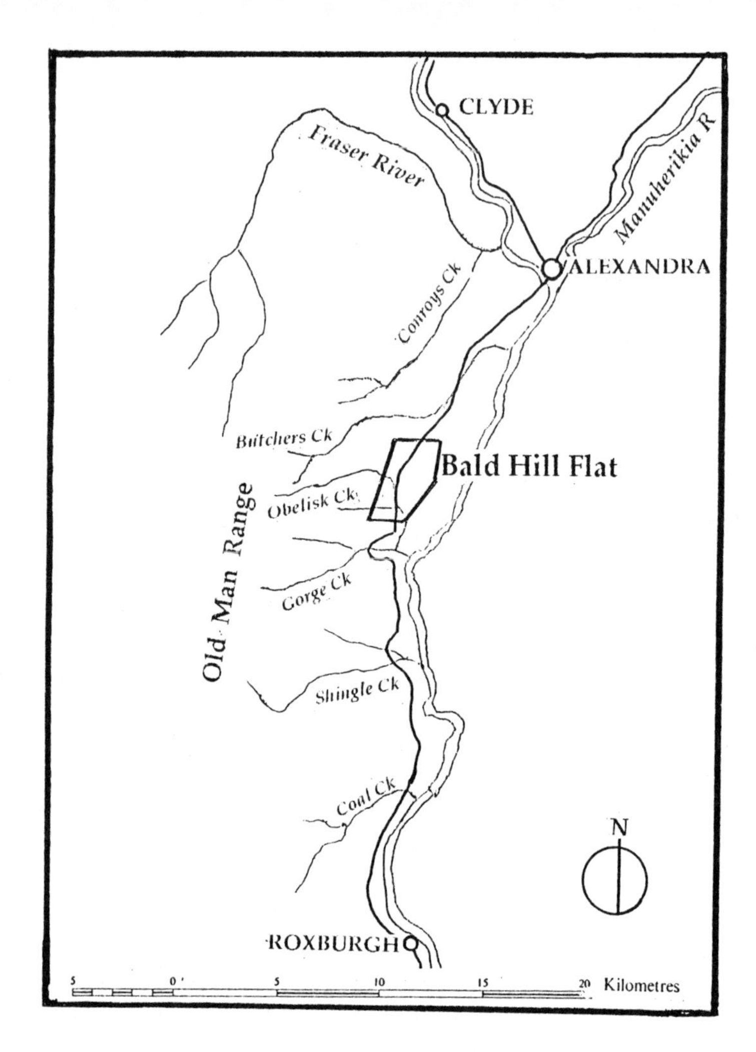

Figure 3.1. Location of Bald Hill Flat (Fruitlands).

It was not until 1915 that the name 'Fruitlands' came into use. It is obviously derived from the name 'Fruitlands Estate' given to the 1,000 acre (400 h) block of land on which the Central Otago Fruitlands Company planted extensive orchards. Although the fruitgrowing project failed, the name has persisted, perhaps because it is more euphonious, or because it is an ironical reminder that it pays to investigate schemes thoroughly before rushing into them.

How Bald Hill Flat was Formed.

Bald Hill Flat forms a broad saddle, about a kilometre wide and four

Figure 3. 2. Bald Hill.

kilometres long lying north to south between Butchers and Obelisk Creeks. It lies at an altitude of some 400 metres on the flanks of the Old Man Range. The western margin of the Flat merges into the slopes of the range, but the eastern margin is more distinct, lying against a steep fault-scarp about 200 metres high. This long scarp bounds a flat-topped hill known nowadays as 'Flat Top Hill,' but in earlier times as the 'Twelve-Mile Range,' which separates Bald Hill Flat from the deep gorge of the Clutha River.

The Flat is a 'fault-angle' valley, and is the result of a giant crack, or fault, which developed in the earth about five million years ago. The fault ran in a north-south direction, and the land on the western side slowly hinged down. This gave rise to an asymmetric valley that sloped gently down from west to east ending at the base of a steep, almost vertical cliff, which is the scarp of the fault. At the time this movement took place, much of the schist rock of Central Otago was covered by clays and sands, which had been deposited on the floors of former extensive lakes. These 'cover beds' subsided with the underlying rock and were, in turn, covered with schist gravel washed down from the slowly rising Old Man Range.

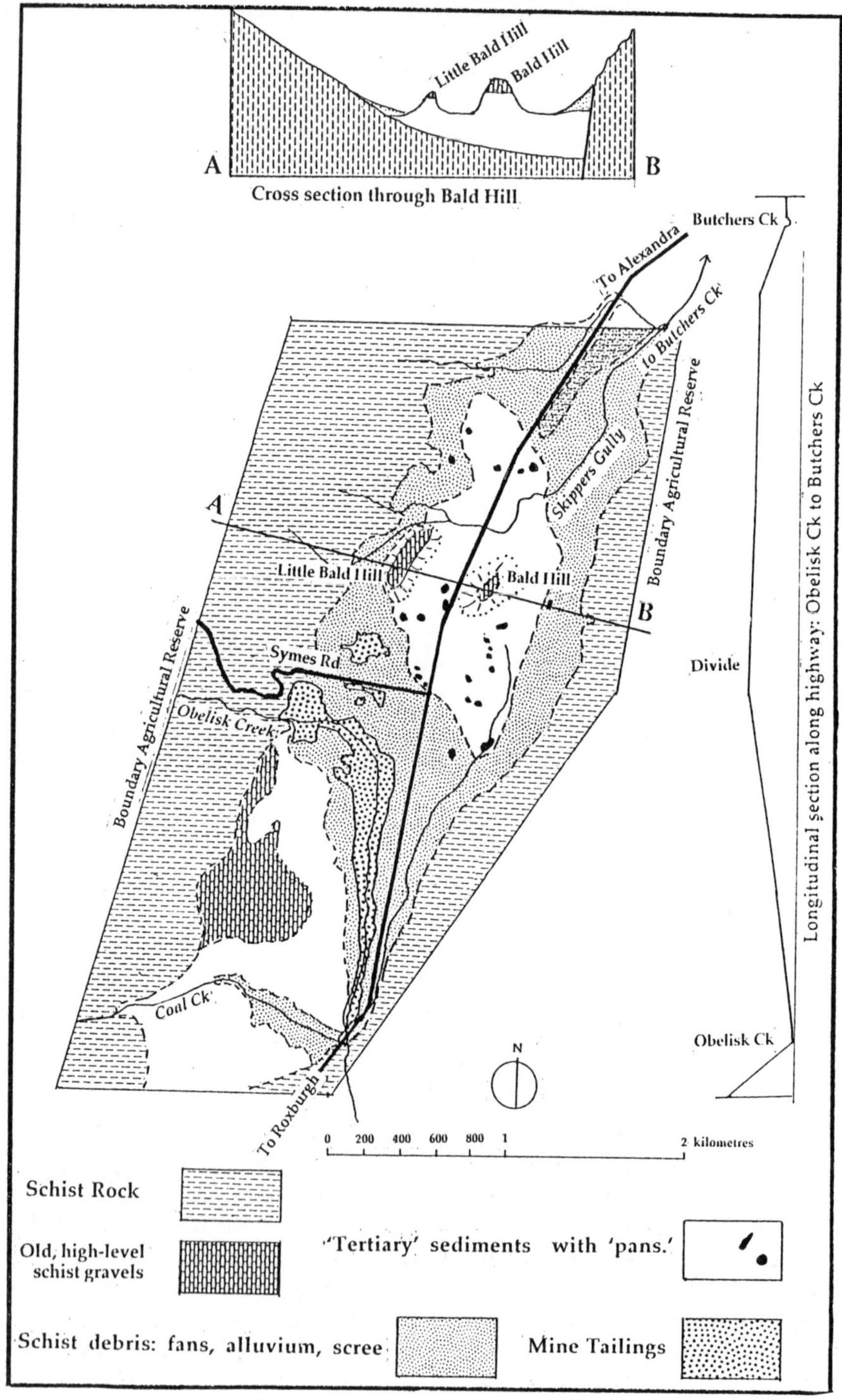

Figure 3. 3. Simplified geological map of Bald Hill Flat with longitudinal and cross sections.

Some of the ephemeral streams flowing from the slopes of the Old Man Range across Bald Hill Flat turned northwards to become tributaries of Butchers Creek, whereas others turned south towards Obelisk Creek. They cut into the cover beds and their overlying gravels, eroding off about 30 metres of the soft materials. A large area of these cover beds still remains between Obelisk and Gorge Creeks at the south end of Bald Bill Flat, but the most spectacular remnants are the two small hills standing on the indistinct, low divide between Butchers and Obelisk Creeks. These isolated remnants are now called 'Bald Hill' and 'Little Bald Hill'. The planed-off expanse of clays and sands was partly covered by fresh fans of schist gravel and sand brought down by the streams flowing from the Old Man Range. The largest of these alluvial fans was deposited by Obelisk Creek, but this creek has now cut a broad course down into its fan and formed a wide flood plain 20 metres or so below the older level. Over about half the area of the Flat, however, the old clays lie close to the surface and their presence is marked by characteristic small, shallow depressions, or 'pans,' which in these days of irrigation, become small ponds.

A feature of Bald Hill Flat which attracted early would-be farmers was the deep, well-structured soil that has developed on the silts and fine sands lying over the schist gravel.

Gold Discovery

When the Clutha River rose in September 1862 and swept the Dunstan Rush miners from the river beaches, they were forced to prospect further afield. It was not long before the wealth of the numerous streams falling from the eastern face of the Old Man Range was uncovered. First it was the fabulous treasure of Conroys Gully, followed quickly by strikes in Butchers Gully and the tributaries of Omeo Creek — Meredith and McLellan Gullies, the miners called them. Every gully along the range was prospected, but it was not until 1 August 1863 that a rush took place to 'Coal Creek' (almost certainly Obelisk Creek[2]), at the southern end of Speargrass Flat.

Early Mining

Here, Andrew Drew and Ben Buchanan had reported they were getting one ounce in 30 buckets from a dry gully where the wash was only a foot (30 cms) below the surface. Soon 80 men were working in two gullies, called Madmans and Drews, and earning 20 to 30 shillings a day. By 1864, a water race had been brought in from further up Obelisk Creek with the intention of sluicing the gullies 'from top to bottom.'

As with most minor gold rushes it was soon over, with miners departing for the latest discovery always hopeful it would be better than the last. A number, though, remained, and over the next few years the whole of Speargrass Flat was prospected. Small amounts of gold at shallow depths

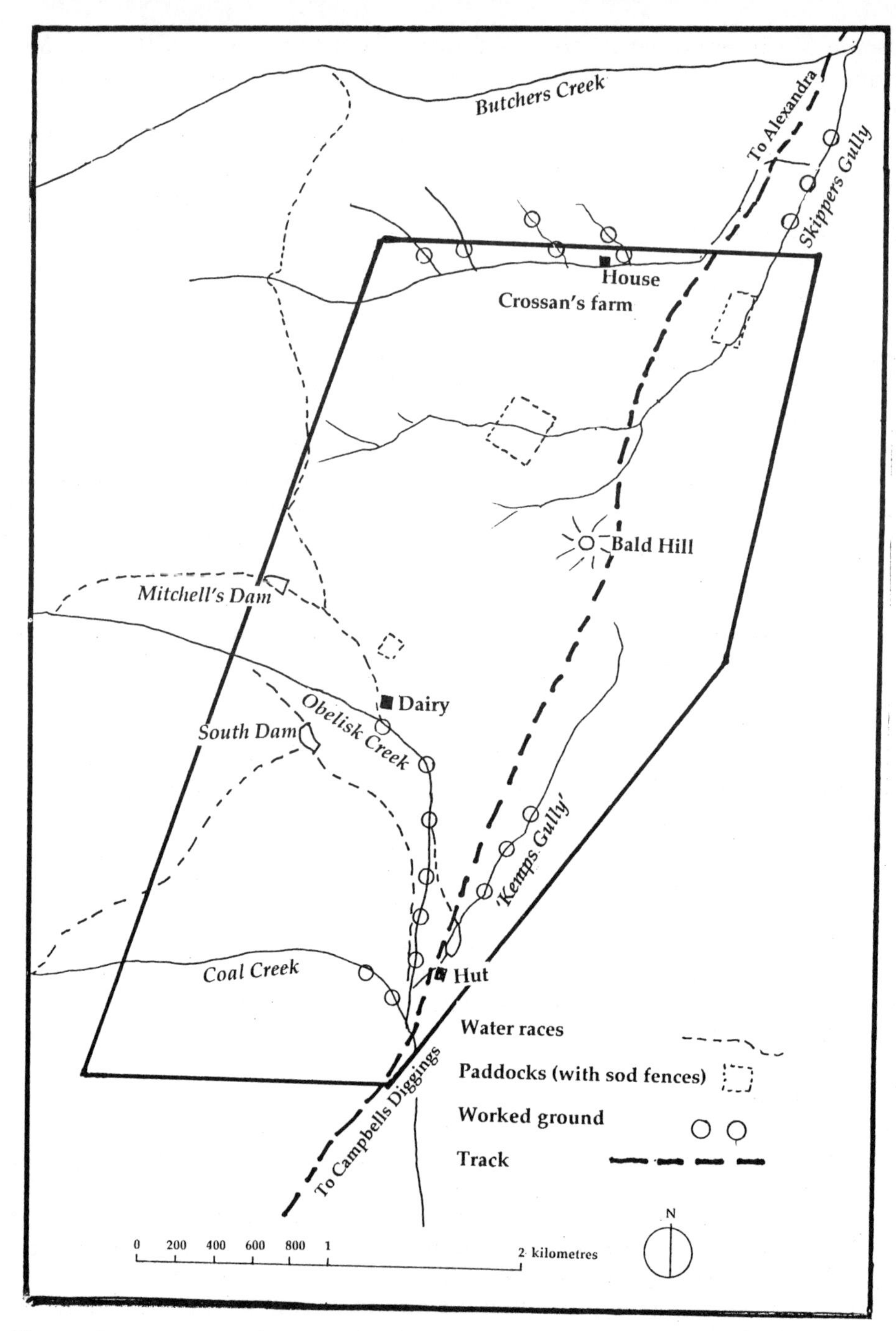

Figure 3. 4. Block of agricultural land surveyed out of the Teviot Run in 1868. Water races, dams, gold workings and other features existing at that time are included.

were found almost everywhere and there were several places that might have been worth sluicing if adequate water had been available.

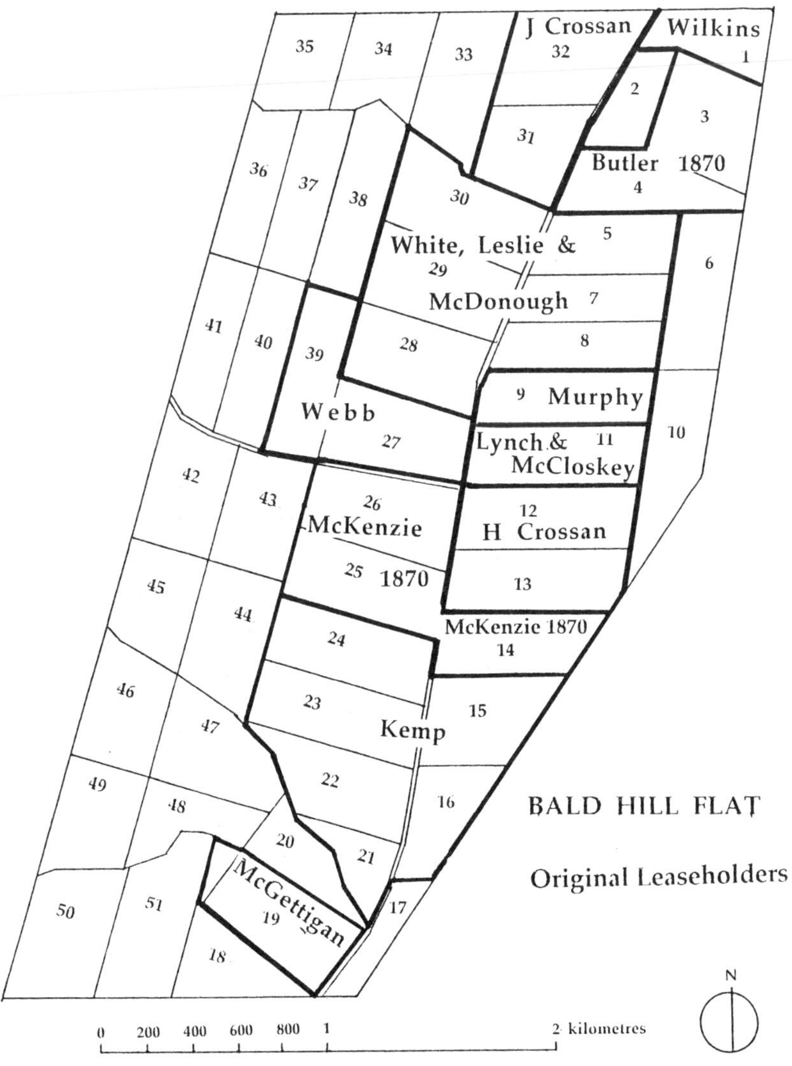

Figure 3. 5. In 1869 Bald Hill Flat was subdivided into 51 sections and within a short time many had been leased to farmers.

These likely prospects included several dry gullies in the rocky hill country along the northern boundary of Bald Hill Flat, immediately adjacent to John Crossan's farm where payable gold had been discovered. A party of six, led by Robert Mitchell, cut a race from Butchers Creek and was able to bring some water to the locality. Soon 30 to 40 men were at work. Then gold was found in the lower reaches of Skippers Gully, formed by a tributary of Butchers Creek, where 60 men were soon panning and cradling. They found gold also in the bed of the small stream (here called 'Kemps Gully') which runs south along the eastern side of the Flat into Obelisk Creek.

These shallow deposits were of small extent and were difficult to work because of lack of water. It did not take the miners long to realise that it was the flood plains of Obelisk Creek, and its tributary Coal Creek, that were likely to provide the only worthwhile mining on the Flat.

Obelisk Creek rises high on the slopes of the Old Man Range, and for much of its course flows through narrow valleys or gorges, but where it crosses Bald Hill Flat it has formed a sloping flood plain 100 to 200 metres wide. It was on this flood plain that mining activity became concentrated.[3]

First Settlers

When John Kemp arrived at Speargrass Flat in 1865, the few people he found settled there included the partners John White, Robert Leslie and John McDonough who ran a dairy farm and sold their produce in Alexandra. Paddocks were demarcated by sod walls, and stock water provided by water races from creeks. The Flat, at this time, was part of the vast Teviot Run (No 199) leased by John Cargill and Edward Anderson, and the early settlers had, at best, only informal arrangements with the runholders to use the land. One at least, John Crossan, is recorded as having the permission of the runholder to settle on the land.[4]

These early settlers had come as miners, but attracted by the fertile soils, found farming gave a more regular income. Their Miners' Rights allowed them to occupy, under certain conditions, a 'Residential Area' of one acre. On this they had to build a dwelling, and so long as they paid the small rent, they had security of tenure. They could also apply for water races from the streams. Kemp himself had a mining claim in Obelisk Creek, but he also grew vegetables on his one acre and sold them to diggers. In addition he packed goods over the Old Man Range for James Rivers who had a branch store at Campbells Gully. But, apart from their 'residential areas,' the settlers had no title to the land they were farming, and understandably, they felt this was an unsatisfactory arrangement.

Finally the settlers met Anderson, the runholder, on the ground to discuss arrangements about their tenure, but he drove too hard a bargain and the meeting broke up. The settlers then enlisted the support of the townspeople of Alexandra and Clyde to petition the Provincial Government. They wanted Bald Hill Flat, as the whole area was becoming known, surveyed off from the Teviot run and subdivided into small holdings.

The Provincial Government was sympathetic, but it was not easy to persuade wealthy and influential runholders to give up part of their land. Eventually, though, 2,046 acres (828 h) which comprised most of Bald Hill Flat, were surveyed off from the run in early 1868. Presumably the runholders had to be compensated for the loss of this attractive block. In 1869 it was subdivided into 51 sections ranging in size from 17 to 50 acres, and immediately settlers applied for Agricultural Leases even before the Proclamation was Gazetted in February 1870.[5] Within a year or two all the land had been taken up, apart from a few sections where miners had successfully objected on the grounds that the land was gold-bearing.

Perhaps the first to take up a lease was J. R. Kemp, who then built a small hotel and store. Later he was able to freehold the land and set about building a much larger two-storied hotel. When it was opened in 1874, this Cape Broom Hotel (named after the 'Cape broom,'[6] a plant used for hedging in the hotel garden) was described as 'without exception the prettiest on the road between Dunedin and Queenstown.'

Miners' Concern

The first water race was cut early in 1864 and soon mining was well established on Speargrass Flat, as the miners referred to the flood plain of Obelisk Creek. It was only under the gravels of this creek that there was sufficient gold to make cradling profitable. Most miners were still cradling but one party at least, Robert Webb and Carl Sorrensen, were ground sluicing on their three acre claim near the junction of Obelisk and Coal Creeks. It is likely that they built the stone-walled dam across a gully south of Obelisk Creek, referred to here as 'South Dam.' Another storage dam ('Mitchell's dam') was built behind the site where the now well-known Mitchell's cottage would be built. The mining population was swelled by those such as George Wilkinson, who had tried the shallow diggings at the north end of the Flat but had decided that Obelisk Creek offered better and more permanent returns. There were sufficient people at Speargrass Flat diggings to support a store, which was owned by Robert Ellison.

With growing alarm the miners watched land which they considered to be 'auriferous,' being leased to the farmers. This in itself did not worry them too much, as the land was still Crown Land and so could be mined with the appropriate licences. The real concern was that the terms of the Act,[7] under which the leases were granted, allowed the farmers to freehold the land after leasing it for three years. When this happened the land would be effectively closed to mining, as it could then be worked only after negotiation with the owner and payment of compensation. It was ironical that this concern over availability of land arose just as a proposal for a new race from Gorge Creek promised abundant water.

Gorge Creek Race

Gambling that the land tenure problem would be solved and that mining would continue, a group announced it planned to bring in 16 heads of water from Gorge Creek to relieve the chronic shortages on Bald Hill Flat.

Miners hoped that the race would supply water to all of the miners on Obelisk Creek. But it was not to be.

The Gorge Creek Water Race Party, as it called itself, included John Mackersey, Henry Forrest, Henry Young, James Simmonds and others. Construction of the race began in 1871 and by January 1873 the water was at Coal Creek. A few months later the race was able to command the mining area of Obelisk Creek and the big race, built at a cost of £3,000, was rapidly nearing completion. But where, asked a newspaper, was the large area of available auriferous ground required to make full use of this amount of water? Nearly every acre was already locked up under the Agriculture Lease system. 'For the sake of a few paltry bushels of oats the interests of our largest producing interests are sacrificed.' it went on.[8]

Miners began to demand that large parts of Bald Hill Flat be reserved for mining. As a first move an Alexandra Miners' Association, headed by Jeremiah Drummey with James Simmonds as secretary, was formed in 1872. It promised to deal with the concerns of all miners, but it was clear that it was set up because of the situation at Bald Hill Flat, and that this issue was going to be its first priority.

Early in 1873, the Association wrote to the Waste Lands Board, and to the Superintendent of the Province, asking that an Inquiry be held into the extent of auriferous land on Bald Hill Flat, and to ask that any such land be reserved for mining.

Little had so far been disclosed about the Gorge Creek Party's plans and the real reasons for building their large race, but eventually it was revealed that the intention was to work a claim on Section 27 near the centre of the Flat. Some mining had already been done on this section and some gold obtained — sufficient to lodge a successful objection and prevent the section being leased to farmers, but the inability to dispose of the tailings had brought mining to a standstill.

The plan to mine Section 27 in the heart of Bald Hill Flat caused great concern to the farmers. who became even more alarmed when a scheme was put forward to build a giant tail race over a mile long. The farmers realised that such a tailrace would allow wholesale mining of much of Bald Hill Flat.

The Sludge Channel

It had been soon realised by the Party that the tail race needed to dispose of the tailings from mining operations on Section 27 would need to be large, very long and expensive. It would cost, in fact, rather more than the Party could afford after building the water race. Then the Party had had a bright idea. The plan was revealed through a series of 'press releases' that would have done credit to a modern public relations expert.

A news item first appeared in March 1873. A proposal had been put forward, it was announced, to construct a giant communal tailrace or 'sludge channel' right into the heart of Bald Hill Flat. It was suggested

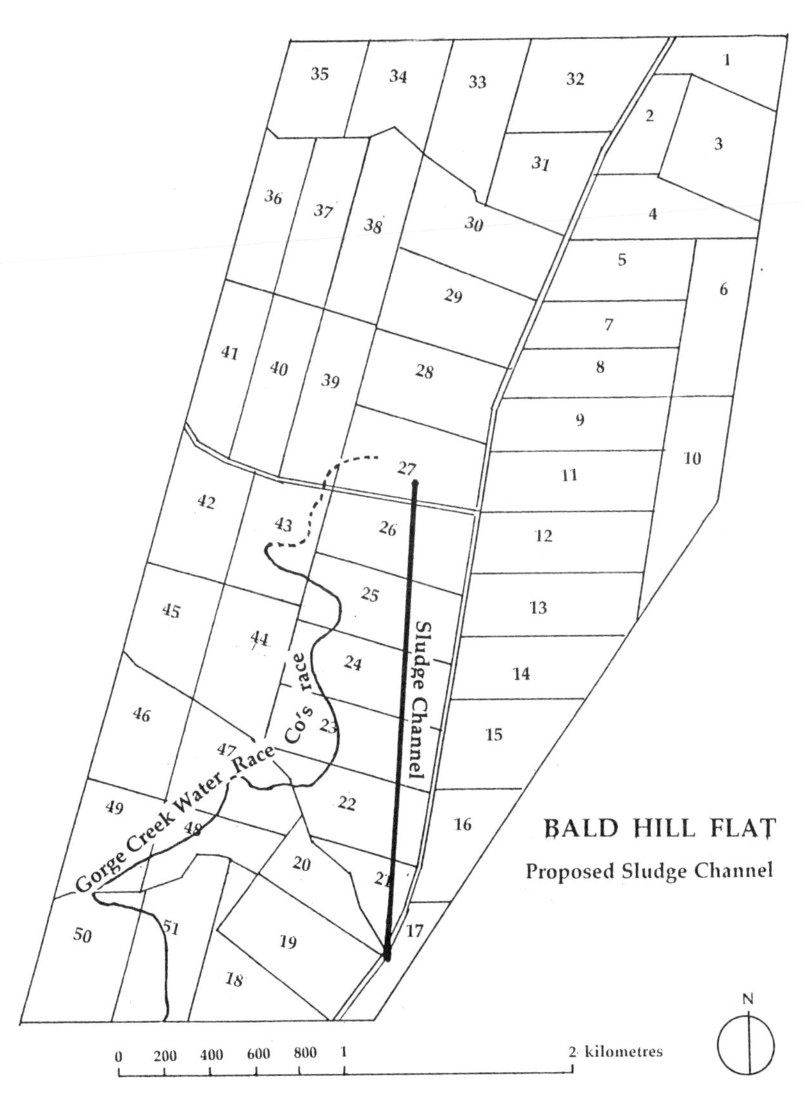

Figure 3. 6. The Gorge Water Race Company's race was intended to sluice Section 27 but was stopped at Obelisk Creek. A long communal tail race (sludge channel) was proposed to deal with tailings.

that such a structure, as a public amenity, would attract a Government subsidy, as had the recently completed tailrace at Mt Ida. [8]

A month later a Letter to the Editor appeared supporting the scheme and suggesting, cunningly, that John Kemp and Hugh Crossan, two large leaseholders, be the promoters. It was also announced that a meeting would be held shortly to form a 'ways and means' committee.

It was this meeting,[9] held in Kemp's Cape Broom hotel and chaired by Crossan, which brought matters to a head, and polarised the community. James Simmonds went into detail about the proposal, and was careful to

Figure 3.7. By 1874 a Mining reserve had been established and a number of other sections had been reserved for mining. Farmers had freeholded the remainder.

emphasise the great benefit that this sludge channel would bestow on the district. It would be taken up through Kemp's and Crossan's properties (they would receive compensation), and continue on to Section 27 in the middle of the Flat. The channel would be over one mile long, and would probably cost about £2,000. It would be paid for by those who would make use of it, buying £5 shares. Simmonds went on to say that it was likely that the whole of Bald Hill Flat was auriferous as every hole dug showed gold, although under questioning, he admitted that very few holes had been dug. He quoted McDonough, who was reputed to have recovered gold at the rate of £10 per week while digging the foundations of his house, and Crossan who found gold while sinking a well.

Battle lines were drawn. The spirited opposition was led by Robert Webb who, as a typical miner/farmer, not only had his own tail race and so wasn't interested in joining the scheme, but also had a leasehold farm next to Section 27. Webb questioned Simmonds closely about the return on the investment that could be expected, but Simmonds was evasive and Webb was able to leave the impression that the main benefactors would be the Simmonds party. Finally Webb moved

. . . knowing that Bald Hill Flat as a whole has been fairly prospected — unless in Section 27, no payable ground has been found — deem it most unwise to construct a sludge channel; and consider the construction of such a channel a waste of labour and money. This motion was put to the meeting and passed by what was described as a 'clique of farmers.'

The battle then moved to the correspondence columns of the newspapers. Simmonds' case wasn't helped by McDonough and Crossan both denying that they had found gold during their excavations. Then others found fault with Simmonds' mathematics by pointing out that there was much less fall along the route of the proposed tailrace than the figure he quoted. In fact, if built to his specifications, the top end of the sludge channel would be 65 feet (20 m) above the ground! It was pointed out by someone else, that part of the reason for pushing the sludge channel was that the Gorge Creek Party would greatly benefit by supplying the large demand for extra water that would undoubtedly be required, not only for mining but also for sluicing the channel itself. Finally they attacked Simmonds personally, pointing out that he had selected information with a bias against the farmers, and that he had little experience in mining compared with those who had lived on the Flat for years as miners — but also, it might have been pointed out, as farmers.

Opposition to the sludge channel was so strong that Simmonds and company were forced to abandon the project. It was obvious, that with the objections they would face, they were unlikely to have the scheme approved by the Warden's Court, especially when it was known that moves were afoot to establish some sort of Mining Reserve.

At a meeting at Bald Hill Flat in early August, the miners decided to put the whole matter of land for miners before a public meeting to be held in Alexandra on 9 August 1873.

At the public meeting[10] it was pointed out that it had been the practice of the Waste Lands Board to grant all leases applied for, unless miners came forward with objections. Only a few sections had escaped being leased. Examples were Sections 1, 2, and 3 at the north end of the Flat, which enclosed a partly worked gully, and Section 27 which was not leased because it had gold workings right in its centre.

Farmers who had held their leases for three years were now wanting to

Figure 3. 8. The Mining Reserve (middle background) about 1912.

purchase the freehold, as they were now perfectly entitled to do under the Act. After all, they would rather be paying off the sections than paying rent. It was just as the miners had predicted and they were worried. In the end the meeting set up a committee to again ask the Waste Lands Board for an Inquiry to look at the problem of mining versus farming land at Bald Hill Flat.

All the farmers were greatly concerned about the rising clamour from the miners, and the meeting at the Cape Broom Hotel had brought things to a head. They realised only too well that a sludge channel, such as described by Simmonds, would enable most of the Flat to be mined. Little wonder that they were against the proposal.

In July 1873, the settlers received word that an Inquiry, asked for by the Miners Association, would be held. They knew that miners such as Forrest and Simmonds were men of influence, and could perhaps persuade an Inquiry that large areas of Bald Hill Flat should be set aside for mining.

Farmers Unite!

The farmers decided that there would be strength in unity. Hurriedly they formed The Settlers' Mutual Protection and Progress Association. The objects were straightforward: to ensure that each settler received the freehold of his leasehold land as soon as possible, and that no miners were allowed on leased land except under the stringent conditions governing mining of leasehold land. At the first meeting on 3 August 1873 the members agreed to draw up an inventory of their land, together with the value of improvements on each block. This was to be the basis for claims for compensation in case parts of the improved farmland were taken into the proposed reserve. At the second meeting a fortnight later, the Association engaged a lawyer to represent their interests at the forthcoming Inquiry.

The Inquiry

The Inquiry[11] was held at Clyde on 27 August 1873 before Lands Officer Mr W. L. Simpson, and Messrs Hazlett and Thompson. Mr W. F. Forrest appeared for the miners and Mr Wilson, a solicitor, represented the leaseholders.

Forrest explained that the miners did not claim that there was rich gold all over Bald Hill Flat, but as well as the clearly auriferous flood plains of Obelisk and Coal Creeks, there were one or two ill-defined gullies which nevertheless contained gold-bearing wash. In particular he drew attention to the branches of Skippers Gully (an ephemeral tributary of Butchers Creek) which crossed the Flat near the Bald Hill itself. Forrest also drew attention to the gully that drained the southern end of the Flat and joined Obelisk Creek near Kemp's place. Here again good prospects had been found.

Evidence was given by miners that gold up to £3 10s a week could be easily recovered from Skippers Gully, and if water had been available, returns could have been as high as £8 per week. There was gold too, in the shallow gullies running down towards John Crossan's land, and there was no reason to think that it would not continue into his section (Section 32).

Other miners testified as to the gold found in Section 27 right in the middle of Bald Hill Flat. In spite of the fact that a mistake was made in levels when a tailrace was constructed into this section, a steady £4 a week could be made. Some days half an ounce of gold was obtained, but shortage of water was the limiting factor.

When the leaseholders gave evidence it was clear that, apart from a few acres on Section 27, very little mining had been done outside the valley of, Obelisk and Coal Creeks. Certainly prospecting holes had been put down on a number of sections in various parts of the Flat, but the fact that they had not been developed into mines told the story.

The Mining Reserve

After some deliberation, the Inquiry recommended that a Mining Reserve be declared, covering mainly the flood plain of Obelisk Creek which had been proven to be auriferous. Other sections reserved for mining, but apparently not included as part of the actual Mining Reserve, included Sections 26 and 27 on the terrace north of Obelisk Creek; Section 1 covering part of Skippers Gully and Sections 42 and 43 in upper Obelisk Creek. The Reserve was surveyed off in 1874, and from 1876 onwards licences for Extended Claims were issued for the Reserve.

One of the licences issued was to the Gorge Creek Company, which had abandoned the idea of the large sludge channel, and with it the original intention of mining Section 27. The big race from Gorge Creek was sold to a partnership of James Hesson, William Lynch, and James Simmonds who took out an eight-acre claim on the Mining Reserve. Although James Simmonds retained a large interest in the concern, he himself, in his restless way, went on to other things.[12]

The battle was over. Both sides had won. The farmers were pleased that they had preserved the bulk of the fertile land of Bald Hill Flat, and the miners had gained the exclusive use of what was widely regarded as the richest gold-bearing ground on the Flat. From this time onwards, mining was confined to the reserved land apart from a few special arrangements made with farmers from time to time, to mine freehold land.

For the next 30 years Bald Hill Flat remained an important alluvial goldfield.

NOTES

1. *Otago Daily Times* 17 January 1911.
2. *Otago Daily Times* 8 and 18 August 1863. 'Coal Creek, so named from the bed of lignite at its head.' (*Otago Daily Times* 23 September 1863). So far as is known, no actual coal has been found in Coal Creek, but according to Mr John Reid, a long time resident, a number of years ago a slip uncovered a band of dark carbonaceous silt which could easily have been regarded as 'coal' by early miners. The site has since been buried again by further landslides.
3. It is possible that it was this flood plain that was called 'Speargrass Flat' and the more northerly part of the Flat (where Butler lived) may have been 'Bald Hill Flat.' For some time both names may have been in use, as the census of 1878 records the population of the two places as separate localities. For a number of years, miners continued to use 'Speargrass Flat' as the location of their claims on Obelisk Creek.
4. Stated at the Inquiry. *Tupeka Times* 4 September 1873.
5. Otago Provincial Government *Gazette* 23 February 1870
6. *Otago Daily Times* 19 December 1874. Kemp's garden was subdivided by hedges of 'Cape broom.' The plant now known as Montpellier broom (*Cytisus monspessulanus*) was originally identified as*Cytisus capensis* (Cape broom) by J. F. Armstrong in 1872. It is softer than common broom and has its flowers in clusters at the ends of branches rather than in the leaf axils.
7. The Otago Waste Land Act 1866. Land could be freeholded at 20s. per acre.
8. *Tuapeka Times* 27 March 1873.
9. *Tuapeka Times* 15 May 1873.

10. *Tuapeka Times* 21 August 1873.
11. *Tuapeka Times* 4 September 1873.
12. Nearly 30 years later Simmonds, as a major shareholder in the Last Chance Hydraulic Elevating Company, finally mined Section 27.

4.

MINING AT BALD HILL FLAT

— Mining Reserve and Elsewhere

The conflict between miners and farmers was largely resolved when the whole of the Obelisk Creek flood plain was proclaimed as a Mining Reserve in 1873. In addition it was agreed that some other sections that had been shown to be auriferous, would not to be leased to farmers and so were, in effect, also reserved for mining. Once the dispute had been settled, alluvial mining at Bald Hill really took off and continued, with a few hiccups, for 30 years.

Mining in the Reserve slackened during the 1880s after James White and Andrew Mitchell struck a reef* on the Old Man Range, and interest and activity swung towards quartz mining.

The 1890s saw the introduction of hydraulic elevators* which, to some extent, obviated the need for deep tailraces, but they required large quantities of water (about 4-6 heads* per elevator) at high pressure. In spite of these restrictions, a fresh burst of activity took place which lasted throughout the decade. But there was no longer a place for the cradle-man on the Reserve — it was now the domain of the big groups, backed by money, who could employ numbers of men and bring in the large quantities of water required for the elevators. Over 100 men were at work on the Flat at this stage.

Water Problems

Mining on Bald Hill Flat was seriously hampered by shortage of water. Obelisk Creek and its tributary, Coal Creek, like other Central Otago mountain streams, became almost dry during the late summer. Again in the winter, when frost locked up the moisture in the high country, their flows fell away. It was only in the spring, when snow on the Old Man Range was melting and to a lesser extent in the autumn, that there was a reasonable flow in the streams. The farmers too, took advantage of their Miners Rights to draw off water from the streams, and quickly developed a network of water races over the Flat to provide water for their stock.

In the early days when ground sluicing* was the preferred method of

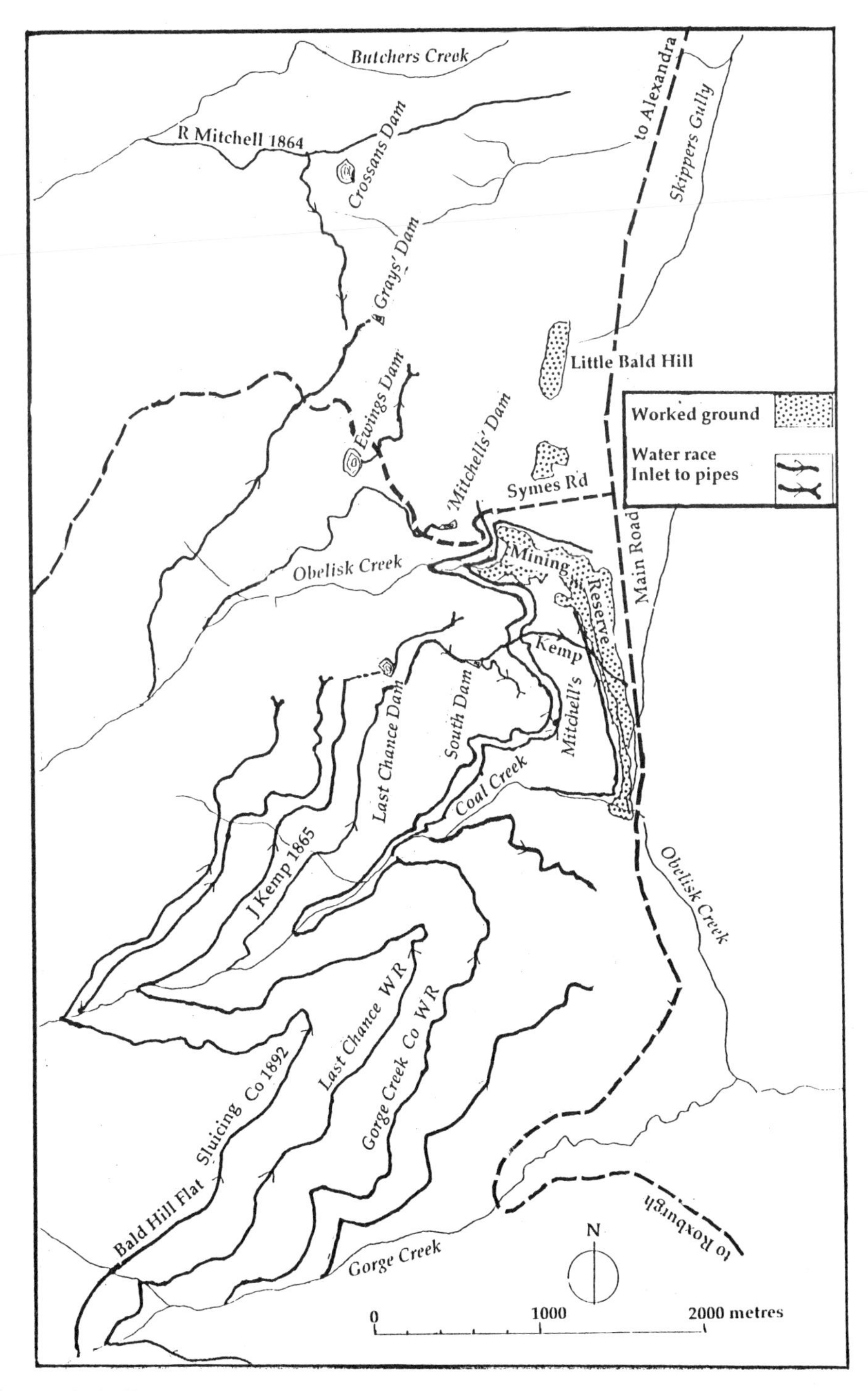

Figure 4. 1. The water races which brought water to Bald Hill Flat. Only the largest can be shown, and some are identified.

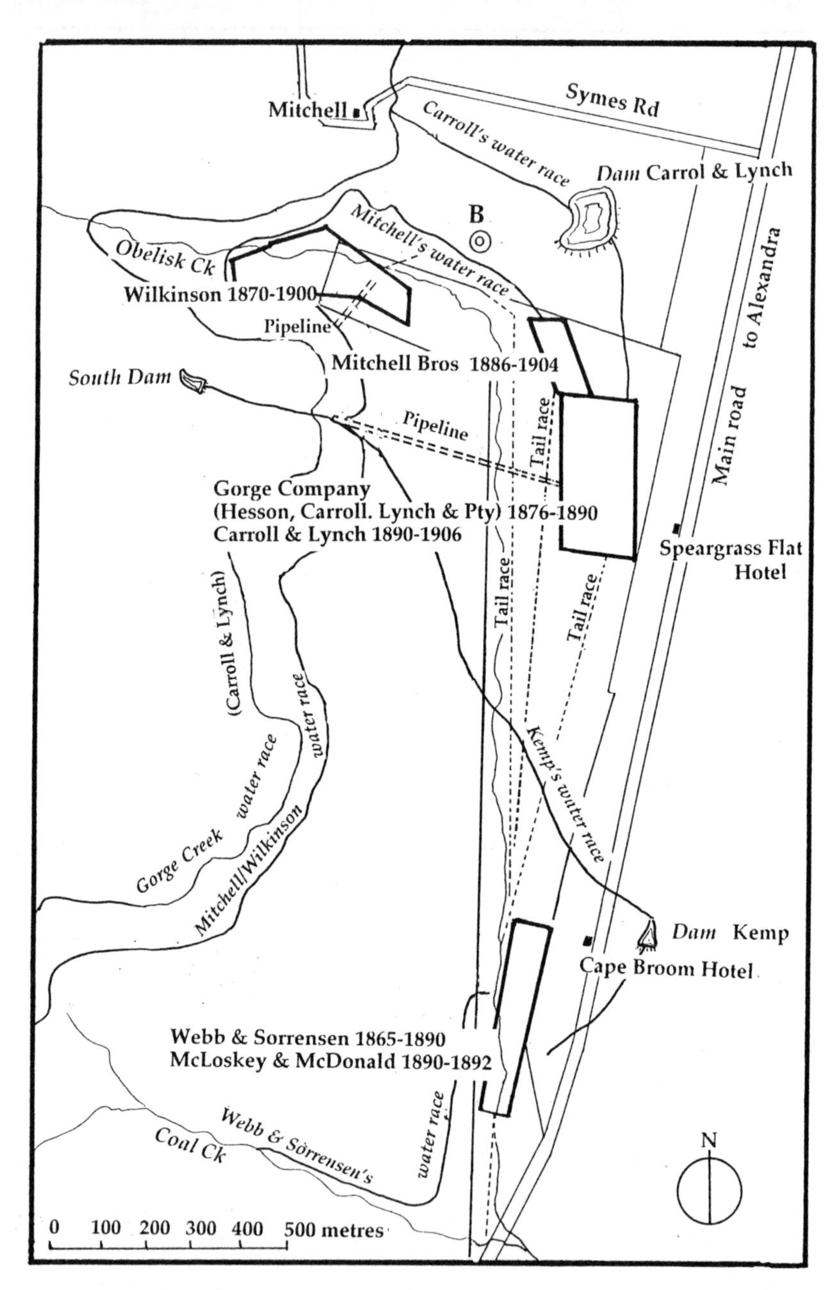

Figure 4. 2. The claims of the smaller partnerships on the Mining Reserve. By 1900 all had been absorbed into larger company holdings.

mining, only short water races were needed to bring water to the claim from higher up the stream. J. R. Kemp, for example, applied in 1865 for a water race only 140 yards long to supply his claim in Obelisk Creek. Soon, however, longer races were necessary. Kemp constructed a long race drawing water from the headwaters of Coal Creek far up on the Old Man Range, and descending by way of various gullies to the 'South Dam.' An even higher race was built by the Old Man Water Company in 1872 to convey water from Gorge Creek across to Obelisk Creek at an altitude of 1,000 metres. This race was eventually acquired by Dr Hyde and at one stage was favoured as a water supply for Alexandra Borough, but was so high that it was buried by snow for part of the year.

The Gorge Water Race Company was formed in 1871 with the intention of constructing a large race from Gorge Creek to Bald Hill Flat. Miners thought that they would be able to obtain water from this race and so end the chronic water shortages. Construction of this major work took place during 1871-73 and was the first of the large races that became a feature of the hillside above Bald Hill Flat.

It was the introduction of hydraulic sluicing* and later, hydraulic elevators, with their demand for large quantities of water, that caused a spate of race building from distant creeks. After the water from Obelisk and Coal Creeks was exhausted, later races were extended to tap Butchers Creek, Gorge Creek, and even Shingle Creek. The large number of traces of old water races leading out of Gorge Creek (seven or eight can be seen in favourable conditions) are reminders of this activity.

With the help of old mining maps, air photos and field inspection, an attempt has been made to map the most prominent races. (Figure 4. 1). A number of indistinct races and the many small distributary races have not been mapped. Where there is no doubt, the races have been identified, but others await positive identification. Confusion in identification can result where the same race is referred to in reports by the name of the miners who happened to be drawing water from it at the time, either by acquiring ownership of the race, by leasing it or by renting the water.

SMALLER PARTNERSHIPS

It is impossible to list all those who mined at Bald Hill Flat, especially in the early days, but as the itinerant gold seekers moved on, a core of permanent miners remained. They formed partnerships of two or more men and sometimes employed others, especially for big 'one-off' jobs such as cutting water races. The partners had often taken up land and only mined when a slackening of farm duties allowed. They built substantial houses, and many spent the remainder of their lives at Bald Hill. Some of their descendants are still there.

Webb and Sorrensen.

Well-established on the Obelisk Creek flood plain as early as 1865, Robert. J. Webb and Carl Sorrensen were perhaps the first to begin ground sluicing at Bald Hill Flat. They worked a three-acre claim near the junction of Coal Creek with water brought in from Coal and Obelisk

Figure 4. 3. The so-called 'South Dam' was probably the first storage dam built for mining purposes in Bald Hill Flat, and dates from about 1865.

Creeks. It is likely that they had a hand in building the 'South Dam' which had been constructed by the time the first survey was done in late 1867. The remains of a stone cottage a few metres downstream from the highway bridge over Obelisk Creek is still pointed out as 'Sorrensen's cottage.'

When Webb committed suicide in the Clutha River in 1890, Sorrensen sold the claim to McDonald and McCloskey[1] who continued working by ground sluicing. Two years later the claim was bought by the Bald Hill Flat Mining Company and incorporated into its larger claim.

Wilkinson

George Wilkinson was another pioneer miner who began before 1870 on a claim at the upper end of the Obelisk Creek flood plain where the creek debouches from a gorge. Wilkinsons's first partner was James Fish, but Andrew Mitchell bought Fish's half share in 1880 for £500 and worked with Wilkinson until 1886. Mitchell then joined his brother, James, in a nearby claim.

Wilkinson was content to work his claim by ground sluicing until late 1892 when he began hydraulic sluicing, resulting in a great improvement in returns, including some 2 oz and 3 oz nuggets. These attracted the attention of the Bald Hill Flat Mining Company Ltd who took up a claim

Figure 4. 4. George Wilkinson mined the upper part of Speargrass Flat for over 30 years. The remains of his house and farm buildings are still visible.

close by. So anxious was this company to get on to Wilkinson's ground that it forced a re-measurement of his claim. This survey found that Wilkinson's claim was indeed larger than the seven acres approved so the extra land was forfeited to the company.

Early maps show a pipeline (labelled 'Wilkinson's pipes') leading from his claim to the water race from Coal Creek labelled variously as 'Mitchell's' or 'Wilkinson's' race. This race was almost certainly constructed by the Mitchell brothers who probably rented water to Wilkinson.

In May 1896 Wilkinson was employing four men working a face five metres high and winning £8 week above expenses. He finally retired and surrendered his claim in November 1900.

A Maori oven, chisel, 'model canoe,' basin and flints had been found in Wilkinson's claim in 1881 at a depth of 10 feet (3 m).[2]

Hesson, Lynch & Simmonds

Frustrated in their endeavours to mine Section 27 in the heart of Bald Hill Flat, some of the partners in the Gorge Water Race Company joined William Lynch, who was already mining on the Reserve. This new partnership, which acquired 8 acres above Webb and Sorrensen's claim, consisted of William Lynch, James Hesson and James Simmonds with Pierce Carroll joining later. In 1876 the partners bought the Gorge Water Race Company's eight-mile water race with its six heads of water from Gorge Creek. Shortly afterwards they began to build a long tailrace which would take many years' work to complete. Later, in 1878 they built a large storage dam on the terrace between their claim and what is now Symes Road.

In January 1890 there was a disagreement between the partners, and as was the practice when disputes arose, the claim, together with the race and water rights from Gorge Creek, were put up for auction. They were bought by two of the partners, Pierce Carroll and William Lynch, for £1,151. Hesson and Simmonds, with other partners, went on to form the Bald Hill Flat Hydraulic Co.

Carroll and Lynch carried on in the original claim. In 1892, after 13 years, they completed their tailrace and were able, at last, to begin recovering gold. They installed an elevator in late 1894 and two years later expanded into a 16-acre claim. Water was conveyed through 2,000 feet (600 m) of pipes from the original Gorge Company water race above their claim. The partners employed between four and six men and won about 1.5 oz of gold each working day. In a good year they recovered about 200 oz.

Mitchell Brothers

Andrew and James Mitchell joined forces in 1886 on a two-acre claim down stream from Wilkinson and just above Carroll and Lynch. They spent four years bringing in a race from Coal Creek, and then in 1888 began work on a 1,900 feet (580 m)-long, perfectly straight tailrace, which was up to 25 feet (8 m) deep. It was finally finished in late 1892. The Mitchells installed a small elevator in mid-1894 but owing to lack of water soon went back to ground sluicing.

Chinese

The Chinese had market gardens and orchards on the banks of Obelisk Creek opposite Wilkinson's claim. Mee Sing bought the gardens but found more profit in sluicing the ground than gardening. His water was from Dr Hyde's race. Ah Gunn had a fruit orchard opposite Wilkinson's claim but also found gold under it and began to sluice it.

Butler's Freehold

Miners could work on freehold land only by arrangement with the land owner. John Butler owned the largest farm on Bald Hill Flat and his big, two-storied stone house and farm buildings were the largest and most elaborate in the district. He allowed miners to work on his land provided they paid £10 per acre and left the land as they found it. Gold had already been found at very shallow depths in Skippers Gully, an almost dry gully which ran across Butler's farm from the lower slopes of the Old Man Range to join Butchers Creek. By mid-1896 more than 20 miners were working in the gully below Butler's house on the east side of the road. The only workings in Skippers Gully on the west side of the road was

Nicholson and Vallance's 'Crow and McDonough' claim.

Others

A number of other small groups worked for short periods on the Mining Reserve. James Halley was working in Obelisk Creek in 1871 and Sam Simmonds and George Rosendale had a claim on the east side of Wilkinson's tailrace in 1875. George Lythgoe and James Gray started in a claim near the Cape Broom hotel in 1876 and began to build a tailrace, but the partnership was dissolved in April 1878. Both men went on to stake out a quartz reef claim on the Old man Range and to form partnerships in several other reef claims. George's brother James was also mining along the creek but sold out in 1890.

THE BIG COMPANIES

Bald Hill Flat Mining Company Ltd,

John Butler also allowed the Bald Hill Flat Mining Company, a Christchurch based public company,[3] to begin mining Little Bald Hill on the western side of the road opposite his homestead. This hill was comprised mainly of ancient clays and sands but was capped by 10 feet (3 m) of schist gravel with about 12 inches of gold-bearing wash at its base. Gold had been discovered here in the very early days but the problem had been how to get water up on to the isolated hill.

Company engineer, B. P. Eckberg, brought water down various gullies from Dr Hyde's race, high up on the Old Man Range, into a storage dam.[4] A short race took the water to a point 150 feet (75 m) above the claim where it entered a 3,600 feet-(1,1000 m) long pipeline which conveyed it across the intervening valley to the claim. The pipes alone cost £1,000. The mine opened in early December 1890 with the festivities usual for such occasions.[5]

The company soon decided that Hyde's race at an altitude of well over 1,000 metres was unsatisfactory as it was under snow in the winter. So it took over Kemp's water rights in Coal Creek and cut a new race, under the direction of the manager, Thomas Donnelly, from Gorge Creek. Water arrived at the claim in mid-October 1892.

Bald Hill Flat Sluicing Company Ltd,

To carry out this expensive race building work the company was restructured as the Bald Hill Flat Sluicing Co Ltd, and took out a £4,000 mortgage with John Ewing, the well-known Central Otago mining entrepreneur.

When Little Bald Hill was worked out, the company intended to sluice Skippers Gully north of Butler's house, but satisfactory arrangements could not be made with Butler, so the company abandoned the project. In January 1894, it moved its operations to the two claims it had acquired in the Mining Reserve.

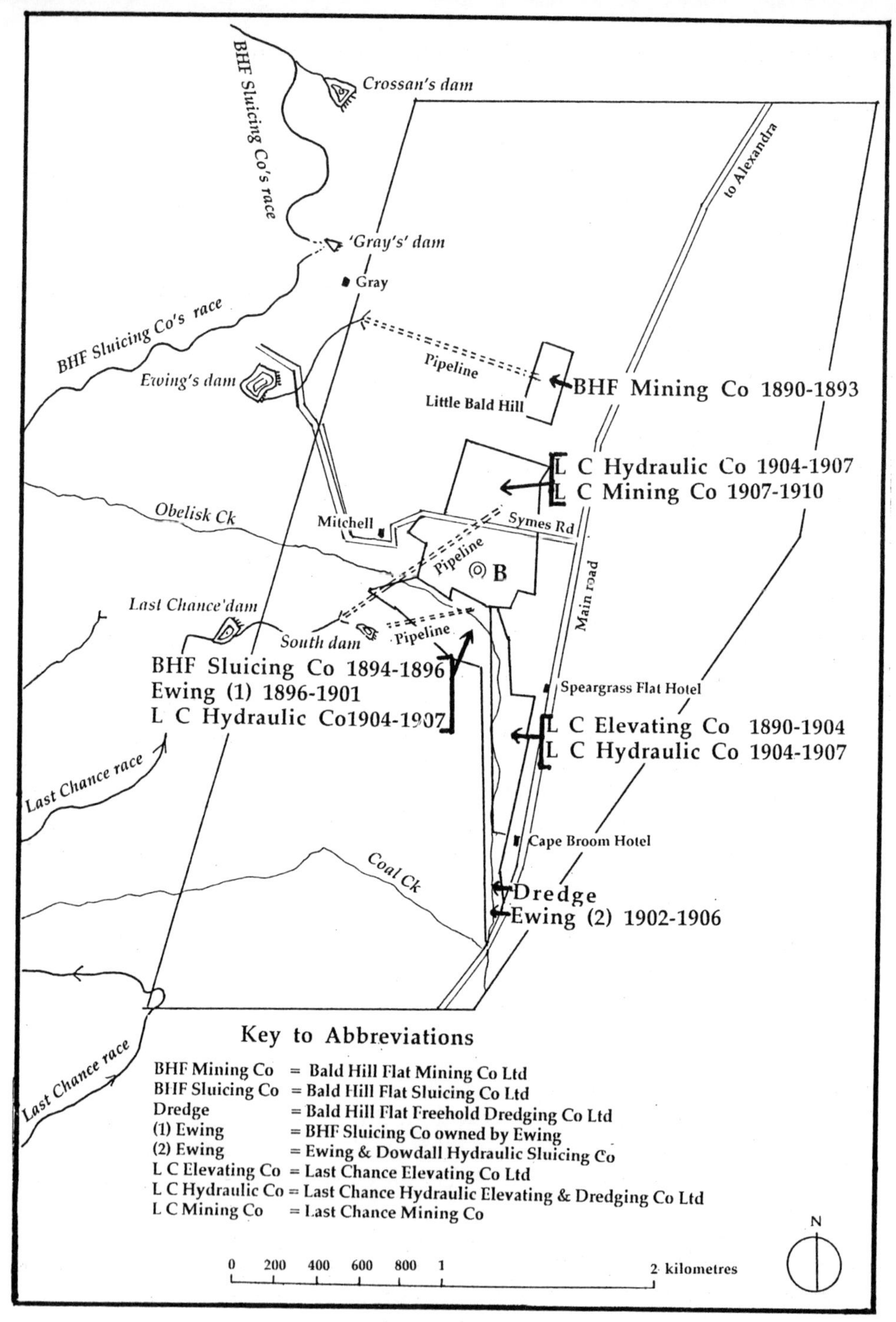

Figure 4. 5. The claims of the companies which worked on Bald Hill Flat. The water races, dams and pipelines are shown where they can be identified.

One, of 11 acres (which included the original Webb claim), was at the south end of the Reserve below Hesson's claim. The other was of 20 acres and adjacent to George Wilkinson's claim where several large nuggets had recently been found. It was late 1893 before the company began laying pipes with a view to setting up an elevator on their lower Reserve claim. But they needed even more water so they bought Foxwell and Party's water right from upper Butchers Creek and surveyed a new race to bring the water to the Mining Reserve.

Figure 4. 6. The remains of Little Bald Hill from the air. Main highway in the background and Bald Hill in top right corner. Just to right of centre in the foreground is a water-filled pan or small shallow formed on Tertiary sediments.

During this change-over period, frustrated and impatient shareholders were no doubt demanding that Thomas Donnelly, the manager, cut down expenses. One of his moves was to reduce the daily rate paid to his workers. Pay for mine labourers varied around the district from 8 shillings to 10 shillings for 8 hours' work — 48 shillings for a six-day week. Donnelly had been paying 8 shillings for work from 7.30 am until 5 pm with an hour off for lunch and two 15 minute spells. The two spells had to be made up, hence the early start. The manager now asked the men to work half an hour longer for the same pay. They refused, so the manager, under instructions from the company, offered them an 8 hour day but for 7 shillings. This caused a strike — a risky procedure on the part of the men as, in the days before unions, this usually resulted in dismissal. But the strike is of interest because the mining correspondent of the *Otago Witness* supported the men and gave the matter some publicity. This allowed a rare glimpse of the working conditions of the time.[8]

As the correspondent pointed out, men were expected to be on the ground well before starting time, already clad in oilskins and gumboots. With the 7.30 am start they would be on the ground close to a total of 10 hours a day, and under the new arrangement it would be even longer. Taking into account the fact that the men were not paid when the mine was not working, which included wet nights, and that the men had to provide their own gumboots and oilskins, it was calculated that they would be lucky to clear 30 shillings a week. The manager defended himself by arguing that 7 shillings or less was commonly paid per day, but it was pointed out that these wages were paid to boys who worked as part of a family concern. It was reported that the matter was settled with the company 'making some concessions.'

The elevator was set up on Webb's old claim and was working by the end of 1894. In the following 18 months the company recovered 500 oz, although it had worked for only 10 months owing to water shortage. But it wasn't sufficient to keep the enterprise afloat.

The company was sold by auction on 3 Jan 1896 and bought by John Ewing (the mortgagee) for £500. Ewing leased Kemp's water for 3-5 years with a purchasing clause. In the end he had six heads of water from Coal and Butchers Creeks. The pressure head was 160 feet (50 m). It is not clear at this stage which dam Ewing used for water storage but it may well have been the old South Dam. The claim covered 37 acres made up of two sections at either end of the Obelisk Creek flood plain connected by a long narrow strip. The claim was abandoned in 1901 and most of it taken over by the newly registered Last Chance Hydraulic Elevating and Dredging Company Ltd.

The Bald Hill Flat Freehold Gold Dredging Co Ltd.

The suggestion that Obelisk Creek might be suitable for dredging led to a spate of letters to the newspapers by 'experts.' Most were of the opinion that the flood plain and the adjoining terraces were very suitable, and mentioned such things as absence of boulders and shallowness of overburden lying over a smooth clay bottom. But none predicted that this clay bottom might be a major difficulty in dredging.

The Bald Hill Flat Freehold Gold Dredging Co Ltd, with a capital of £11,000, was registered in July 1899. The company bought all of James Kemp's and George Burton's freehold land, which included both the Speargrass Flat and Cape Broom Hotels, a total of 300 acres. The men behind the scheme included: J A Chapman (Chairman), J. G. Closs, R. Gunion (Chairman), John Mackersey W. W. Shelmendine and James Simmonds.

Figure 4. 7. The much vaunted dredging venture was an abysmal failure

A large dredge was built and started work in March 1901, behind and just downstream from Kemp's Cape Broom Hotel. Three months later, in June, it had closed down. The reason given was that severe frost made it impossible to work, but there were other reasons. The company had overspent and was in debt. This was not helped by the fact that the dredge had been built in a dam that proved too small. It had to be enlarged before the dredge could move and this delay was expensive. But the most important factor was that when the dredge buckets brought up the gold, it was stuck to clay scraped from the 'bottom' and the two could not be separated easily.

The company had applied for a 100-acre claim but by the time it was processed the company had been liquidated. The dredge was sold to the builder, J Sparrow, the following month for £1,200. It was dismantled in April 1902 and sold to the Parapara claim on the West Coast.

Ewing and Dowdall Hydraulic Sluicing Company

John Dowdall acquired the 300 acres belonging to the failed dredging company, and joined John Ewing in a partnership known as the Ewing and Dowdall Hydraulic Sluicing Company, to mine the land where the dredge had started. W. McNeish was appointed manager and an elevator was used. They were able to demonstrate that in Obelisk Creek, sluicing was a much more profitable method of recovering gold than dredging.

Figure 4. 8. John Ewing inspecting the workings of the claim he held with Dowdall.

When the claim was worked out and abandoned in 1906, the elevator was sold to the Fraser Basin Company and carted into their claim by bullock wagon.

Bald Hill Flat Hydraulic Mining Company. 1890-92

Meanwhile James Hesson and James Simmonds, after selling the Gorge Company's claim and race to former partners Carroll and Lynch, formed the Bald Hill Flat Hydraulic Mining Company with new partners Samuel Simmonds, Peter Jacobs and C. Leijon.

The company was granted 27 acres lying between the claim of Carrol and Lynch, and that of Webb and Sorrensen. Tenders were called for construction of a new six miles (6 km) long race to Gorge Creek at a cost of about £650. Then a large dam was built up on the range 300 feet in altitude above the claim, which together with a pipeline 3,000 feet (900 m) long and a tailrace cost £2,000. By 1891 they had the first elevator in the district installed at a cost of £3,000. To help with the finances Stephen Foxwell bought a 1/6 share.

In March 1892 the company bought the Shingle Creek water rights from the Molyneux Hydraulic Co for £610. Construction then started on a seven mile (11 km) extension to bring the water from Shingle Creek into the company's existing race out of the South Branch of Gorge Creek. Up to 40 men were employed on this work.

Figure 4. 9. The dam of the Last Chance Company was high above Bald Hill Flat. (State Highway 8 can be seen centre right).

At the same time J. Hesson, J. Simmonds, Foxwell and Leijon, using the name 'Perseverance Hydraulic Company,' were granted 40 acres outside the Mining Reserve on the terrace north of Symes Road. This included Section 27, the proposed mining of which had caused all the fuss back in 1873. The forming of this new company with different partners, was a device to acquire even more land and prevent other miners from applying for it.

This two-name arrangement proved to be cumbersome, and in addition, as the original company name 'Bald Hill Hydraulic Mining Company' was being constantly confused with that of the Bald Hill Mining Company Ltd, it was decided to trade under one new name. So from August 1892 the two companies were regarded as one and called the 'Last Chance Hydraulic Elevating Company.'

Last Chance Hydraulic Elevating Co 1892-1899

Quayle's water rights of 12 heads from Shingle and Chasm Creeks were bought for £500 and a large number of men were employed in building a new race from Shingle Creek to Gorge Creek. The water arrived at the claim in March 1893. The company, which now had control of all the water from these creeks, was able to install a second elevator in September 1893.

The company was now operating in a big way. It had 16 miles (27 km)

Figure 4.10. Visitors' day at the Last Chance claim. Note the twin elevators.

Figure 4. 11. An hydraulic elevator hole north of Symes Road marks the last site worked by the Last Chance Company.

of race carrying 10-12 heads (it needed 4-6 heads to work each elevator) which required the attention of several men. Maintenance of the race was difficult and expensive as exemplified in 1896 when a slip swept away 150 yards of the race.

Gravel was sluiced from the seven metre-high face into the elevators, which lifted it 25 feet (8 m) into iron sluice boxes 40 feet (12 m) long fitted with iron ripples, where the gold was trapped.

The mining season was restricted, because sufficient water to work the elevators was only available from the time of snow melting in the late spring until January. Nevertheless substantial amounts of gold were recovered. About 100 oz were recovered each month, with 150 oz recovered for the six weeks work ending 30 October 1896. Ten men, including three of the partners, were employed on the claim with J. McNeish as manager.

In 1896 expansion took place. As well as the 27-acre section in the Mining Reserve, and the 40 acres north of Symes Road, the company was granted 34 acres in the block between the Reserve and Symes Rd.

The Last Chance Hydraulic Elevating and Dredging Co. Ltd 1899-1907.

This public company was registered in July 1899 with a capital of £12,000 and 103 shareholders, with the intention of dredging the claims. Some equipment was ordered but fortunately it was decided to wait to see what success the neighbouring Bald Hill Freehold Dredging Company had with their proposed dredge before finally agreeing to go ahead and order a dredge. Needless to say, the fate of the Bald Hill dredge caused any idea of dredging to be abandoned. Meanwhile the company still sluiced and elevated profitably with seven heads of water, and began paying dividends in June 1903.

By 1904 the company shifted operations from Obelisk Creek to the block it held north of Symes Road behind the school. A trial with the elevator did not pay expenses so the company went back to ground sluicing. But the company had passed its peak, and in September 1907 it was bought by a local syndicate for £450, which cleared its debts, and then the company was liquidated.

Last Chance Mining Co. 1907-1910.

This new smaller company began with six men and W. Neish Snr as manager but in 1910 it petered out. In July 1911 Charles Weaver bought the water rights and later sold them to the Government to form the basis of the Last Chance Irrigation Scheme.

NOTES

1. Spelt 'McCloskey' in the Electoral Roll but mostly 'McCluskey' in the newspapers.

2. *Dunstan Times* 30 June 1882. The 'model canoe' according to Jill Hamel, 1995, would be the hollowed out board used for making fire— a *kaunoti.*

3. This was the fist public mining company to work on Bald Hill Flat.

4. This dam is still used as part of the Last Chance irrigation scheme and is locally known as 'Symes Dam.'

5. *Dunstan Times* 5 December 1890.

6. *Otago Witness* 19 April 1894; 3 and 10 May 1894

5.

EXERCISES IN FUTILITY

Blasting the Molyneux Falls

At Alexandra, the Clutha River turns almost through a right angle and virtually disappears into the mountains as it enters the 30 kilometre-long gorge between the Old Man Range to the west and the Knobby Range to the east. For many kilometres, the gorge is more than 400 metres deep and so narrow that the swirling river at the bottom leaves only enough room for the narrowest of tracks between the water and the precipitous walls of the gorge. Nevertheless, tracks on both sides of the river were soon developed by the foot traffic through the gorge. There was quite a large population of miners, panning and cradling the sands of the river beaches, who were constantly moving up and down the gorge looking for better sites or visiting Alexandra for supplies and entertainment.

Today the gently flowing river, slowed by Roxburgh Dam at the southern end of the gorge, gives little idea of the power and ferocity displayed by the water before the dam calmed its turbulence. It did not have a smooth passage through the gorge. In a number of places obstructions interfered with the flow, but their only effect was to emphasise the power of the current. It crashed through rapids, one of which was so steep that it was referred to as a waterfall, or raced at speeds estimated at more than 40 kilometres per hour through sections where the flow was confined between narrow walls. It was when the river was at its lowest during the winter that its display of power was most evident, for it was then that rapids and falls were exposed. When the flow was high or in flood during the summer the obstructions were completely submerged.

Gates of the Gorge

The first feature of note, just a few hundred metres down the gorge from Alexandra, was a huge, rocky protuberance jutting out from the eastern bank that confined the river to a width of less than 100 metres. Little wonder this narrow entrance was called the 'Gates of the Gorge' by the miners. The river then broadened and flowed serenely, but swiftly, to a

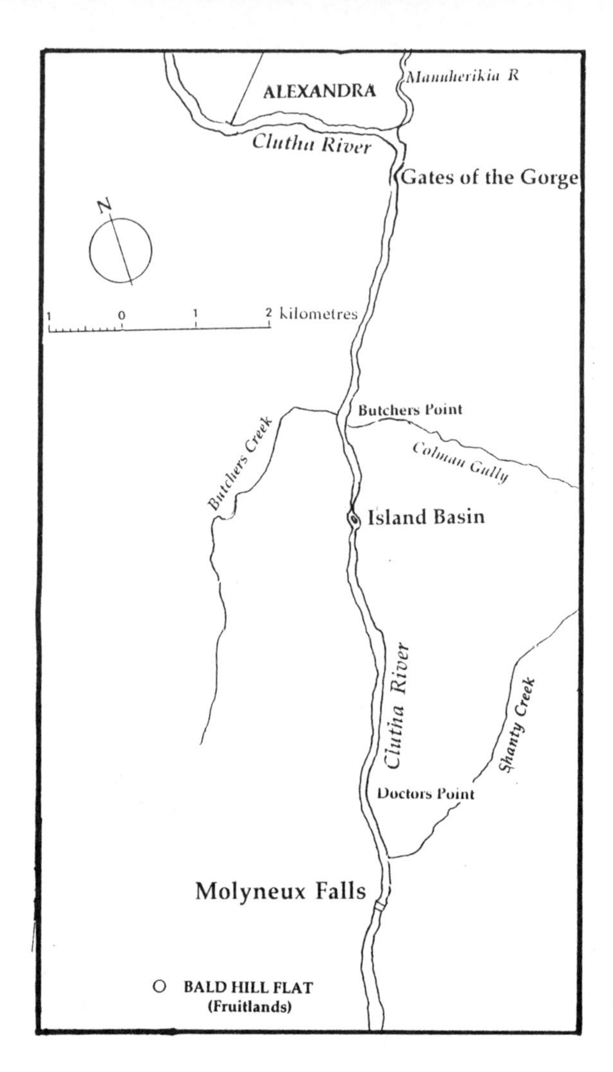

Figure 5. 1. Location map of gorge below Alexandra showing Molyneux Falls and other features.

point nearly two kilometres beyond Butchers Point where it was partly obstructed by the debris of a giant landslide from the left bank.

The Island Basin

A huge area of shattered schist had slumped 200 metres downwards leaving an amphitheatre-shaped scar in the gorge wall, with an almost vertical cliff some 100 metres high as a backwall. The toe of the slump dammed the river which, no doubt, had backed up for several kilometres. When the river overtopped the obstruction and began to cut a narrow channel down through the temporary dam, it washed away the smaller boulders but could not move the big slabs of schist that were in the

Figure 5. 2. An 1865 photograph of the junction of the Manuherikia and Clutha Rivers shows the latter entering the gorge through the 'Gates of the Gorge.'

slumped material. The combined effect of the water racing through the narrow gap and the turbulence caused by the remaining large rocks gave rise to a fearsome set of rapids. The turbulence of the crashing waters scoured out the gravels of the riverbed below the rapids forming a wide basin. The stirred-up gravels were heaped up into a shoal in the middle of the basin. The miners named the place 'the Island Basin.'

Molyneux Falls

Just below the confluence of Shanty Creek was another obstruction also caused by rocks that fell from the gorge walls. This time it was a big rock-fall rather than a slump, but the effect was similar. The river was again temporarily blocked, and again the river managed to clear away most of the debris, but remaining were several huge slabs of rock which the river, for all its power, could not move. It flowed over the obstruction as a steep rapid or waterfall. The miners named the feature the 'Molyneux Falls'. When the river was low it was reported to fall about 12 feet (3.5 m) in a distance of one chain (20 m). During average flow the fall was about four feet (1.2 m), but in flood the fall was submerged and there was little sign of it.

Figure 5. 3. A spectacular slump from the eastern wall of the gorge partly blocked the river. The resulting rapids caused the excavation of the Island Basin with its central shoal of alluvium.

Further down the river there were other rapids, some long and dangerous, but it was the Molyneux Falls that interested the early miners.

As the autumn of 1864 wore on, the river level began to drop as frost locked up moisture in the headwaters. It looked as if it might fall as low as it did in the year of the Dunstan Rush. Crowds of miners began to congregate in Alexandra waiting for their favourite beaches to become exposed.

Then someone had a bright idea: if the obstruction giving rise to the Molyneux Falls were removed, the river would be lowered considerably for four or five miles upstream. River beaches with their untapped gold would be exposed for mining. The proposal began to firm up, and soon over £100 was collected with more promised if needed. The mining surveyor, Mr Coates, offered his expertise for the project.[1]

Early Attempts at Removal 1864

The main obstruction was a huge slab of schist rock estimated to be 100 feet long and 40 feet wide (30 x 12 m) and of unknown thickness. It lay

Figure 5. 4. The Island Basin and its rapids.

obliquely across the river and dipped into the river at angle 45.⁰. The plan was to blow off the end in the water with explosives and then place explosives under the upper part of the rock in the hope that the rock could be thrown bodily over into deep water. If successful the river would be lowered for about 4 miles upstream.

A manager, Mr Ward, was appointed by 1 July 1864 and the miners built a stone bridge out from the bank to the large rock in the centre of the river. The whole project was fraught with danger.

The first attempt to destroy the falls was made by Ward a fortnight later, but it ended, according to the newspaper, 'in ridiculous failure.' Miners, angry over the waste of two hundredweight (100 kg) of powder, took the job out of Ward's hands declaring they could do better themselves.[2] The following week a marine boring apparatus, sent up by the Provincial Government, arrived complete with the necessary drills, eight or ten charges of powder in hermetically sealed tubes and a voltaic battery for firing the charges. All this gear was transported down into the gorge.

Holes were drilled in the big rock and by late July all was ready. The explosives were fired, but either through wrong placing or lack of sufficient charges, the big rock was simply cracked into several pieces, and wasn't moved an inch. The paper commented that it was a pity that the Government had not sent up a properly qualified man to superintend the work and so put a stop to the expense and waste of time.

And that was the end of early attempts to remove the falls. They were left in peace for about 35 years.

Renewed Attempts 1897-1903

By 1897 the gold dredging era was beginning in earnest and dredges were already at work in the gorge, with others planned or under construction. However, the bucket ladders fitted to the dredges that were being built at the time were not long enough to reach the bottom of the river in the gorge unless the river was exceptionally low. A party of Alexandra people proposed to remove the rocks forming the Molyneux Falls and then to dredge the lowered river above the falls. But this scheme, like many other mining proposals, faded away without any action being undertaken.

Two years later the removal of the falls was in the news again, but this time it was not directly related to recovering gold from a lowered river level. It was because the falls prevented movement of dredges up and down the gorge. Certainly two small dredges, while still at the pontoon stage of construction and before machinery had been fitted, had been successfully taken over the falls at high water but no one was prepared to try it with a full-sized, complete dredge.

The presence of the falls meant that the gorge was divided into two parts. Above the falls dredges could be floated down to their claims after being built at Alexandra. Their coal and other supplies could also, with some difficulty, be supplied by boat from Alexandra. But there was no way a big dredge could travel from Alexandra to claims below the falls. This meant that dredges intended to work below the falls, such as the *Fourteen-Mile Beach* and the *Sixteen-Mile*, had to be built in the gorge below the falls. This required extraordinary efforts to land materials at the construction sites. Eventually a track of sorts was found by which a horse-drawn sledge could zigzag its way down the steep wall of the gorge. Heavy components were lowered down the cliffs by ropes.

During a visit to Alexandra in early 1899 by the Prime Minister, Richard Seddon, a deputation from the Central Dredge-owners Association approached him with a petition. It asked that an expert from the Defence Department should be sent to the Clutha River to instruct dredgemen in the use of gun cotton for blasting rocks, particularly the falls. Seddon thought the request was reasonable and he would recommend it to the Minister of Defence, provided that the man's expenses were met. However, before any blasting was done he wanted a report as to what effect the removal of rocks would have on the river lower down.

Inspector of Mines, Mr Hayes duly inspected the falls and could not see any difficulty in removing the obstruction during the months of June to August when the river was at its lowest. He did not think that removing the obstruction would affect any land lower down the river, but he did not not think the work was of national importance as the Dredge-owners

Figure 5. 5. The Molyneux Falls at about mid-flow. At low river the water fell about 4 metres and during floods the falls disappeared.

Association had asserted. He felt that those who would receive immediate benefit should bear half the cost that he estimated at £300.[4]

Although Hayes's report was favourable, Seddon was slow to give a firm answer to the Association's request for expert help. It was only after Ben Naylor and Dr Hyde of Clyde had telegraphed the Premier seeking a response that Seddon replied. He could see difficulties with the scheme, he said. He was concerned that when the falls were removed, the rush of water would sweep downstream gravel and gold-bearing wash from the riverbed above the falls. This might lay those responsible for the removal of the falls liable to upstream claim-holders for damages. Similarly those landholders downstream might bring action for damages through the river level rising because of the deposition of material from above the falls.

Seddon said he had placed the whole matter before the law officials, but it was obvious that he was backing away from his first encouraging response to the Alexandra deputation.

A newspaper editorial could not make sense of this reply. It pointed out that the nearest landholders downstream were at Coal Creek 12 miles (20 km) away, and anyway there was the chain (20 m) wide riverbank reserve along both banks of the river which would allow a huge rise in river level before it affected landholders. So far as the upstream claim-holders were concerned, every one of them was in favour of lowering the river as it could not help but benefit them. The river was at its lowest at the moment and the paper hoped that the Premier would authorise the job as soon as possible.[5]

However, it was not to be. In the face of possible legal consequences, Seddon decided it was not a job for the Government to become involved in. And the estimated cost of £300 was apparently too much for any dredge company to consider taking on.

Eventually practically every metre of the gorge was pegged out for dredging claims. The Golden Falls Dredging Company's claim lay just above the falls, and it was realised that supplying coal to a steam dredge more than six miles (10km) down the river would be a very difficult and expensive undertaking. So it was decided to make use of old technology

Figure 5. 6. The *Golden Falls* dredge was driven by two huge current-wheels and blamed the diminution of current, caused by the obstruction of the falls, for its failure.

and equip the proposed dredge with wheels to be turned by the current of the river. It was calculated that the wheels would provide adequate power to work the dredge and that £50 a week would be saved in operating costs over a steam dredge.

When the *Golden Falls* finally reached her claim in mid-July it was to find that the current was insufficient to generate enough power to work the machinery satisfactorily. What had been overlooked was that the obstruction in the river that gave rise to the falls also slowed the current above the falls.

Not one flake of gold was recovered. Within two months of its arrival at the claim the dredge was sold to Dr Hyde for £1,000. A new company, the Molyneux Falls Company, was registered on 12 February 1903, and the first thing it did was to attempt to get rid of the waterfall at the lower end of the claim. It was thought that if it were removed the current would be speeded up and the dredge might then be successful. So nine hundredweight of gelignite, worth over £70, was exploded, but with little effect on the waterfall.

That was the last attempt to remove the falls by blasting. But when the sluice gates of the big Roxburgh Dam were closed in July 1956, the rapidly rising lake water managed to accomplish in less than 24 hours what much effort over nearly 100 years had failed to do—destroy the Molyneux Falls.

NOTES

1. *Otago Daily Times* 2 May 1864.
2. *Otago Daily Times* 20 July 1864.
3. *Otago Daily Times* 5 August 1864.
4. *New Zealand Mines Record* 16 July 1899.
5. *Dunstan Times* 9 June 1899.

6.

GOLD IN THE ROCK

Quartz Mining

There was something about a gold-bearing quartz reef* that set it apart from other sources of gold in the minds of the public and investors. Perhaps it was the reports of huge amounts of gold won from the famous quartz mines of California, Australia and South Africa. Perhaps it was the perceived permanence of quartz mining as compared with the more ephemeral alluvial mining. Or perhaps it was just the romantic thoughts of dark tunnels with gold shining from the wet rock surfaces. Whatever the attraction, miners were always on the lookout for signs of a reef, especially in the headwaters of the gold-rich streams falling from the eastern slopes of the Old Man Range. It was believed that somewhere up there lay the 'mother lode' — the mythical rich lode that was supposed to be the source of the gold in the streams. So when something that looked like a reef was found, there were plenty who were ready to peg it out, form a syndicate, and begin sinking a shaft, often on only the slimmest evidence of the presence of gold.

Reef mining was not for the individual. Sinking shafts down through rock required explosives, winches, pumps and other expensive equipment. If the initial exploratory shaft showed that the gold continued at depth and the mine was worth developing, then a stamper-battery* to crush the gold-bearing quartz would have to be obtained. A water wheel to drive the stamper was needed, and this in turn required water races with a plentiful supply of water. Then there was the equipment needed to save the gold by amalgamating it with mercury, and then to recover the gold from the amalgam.

The Alexandra district was envious of places such as Skippers Creek, Carrick Ranges and Bendigo in the Upper Clutha valley where fabulously rich lodes* had been discovered. So it is hardly surprising that the reported finding of a reef in the local district, no matter how vague the information, was always given prominence in the newspapers. Each discovery reinforced the popular prediction that the Old Man Range was packed full of gold-bearing reefs just waiting to be discovered, and miners were constantly extolled to renew their efforts to locate them.

In fact, as more and more reefs, or what were thought to be reefs, were discovered along the foothills of the Old Man Range, it began to look as if the predictions were correct. Reefs were reported from the vicinity of Butchers Gully, from Conroys Gully, Blackmans Gully, the head of Aldinga Creek, and even from the other side of the range at Campbells

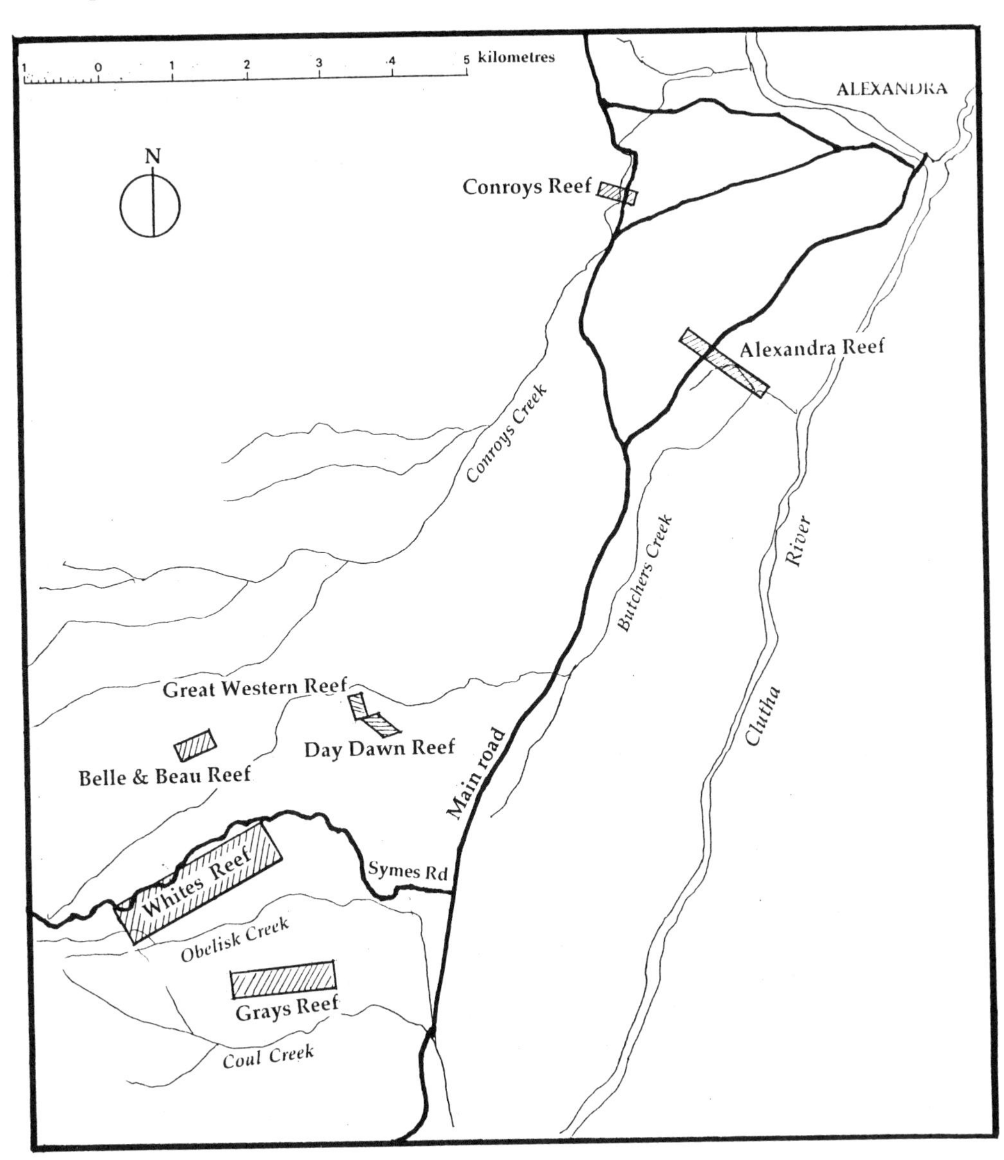

Figure 6. 1. Location of reefs in the Alexandra district.

Gully. Most when tested, were found wanting. It wasn't until the opening up of White's reef far up on the slopes of the Old Man Range above Bald Hill Flat, that it was realised that, at last, a real reef capable of sustained gold production, had been found.

BUTCHERS GULLY REEFS

It seems that the quartz reef Charles Goltz found on the ridge between Chapmans and Butchers Creeks[1] in March 1866 was the first to be discovered in the district. As was often the case, the reef was reported to be 3 feet wide and rich in gold. Further investigation disclosed that the gold was confined to a few narrow inch-wide 'stringers,' and much effort was expended in searching for the supposed 'main reef.'

Edmund Jones, a local mining entrepreneur, financed the sinking of a shaft to a depth of 50 feet (16 m), but soon after it was completed Goltz fell to the bottom and was permanently crippled. Shortly afterwards Jones died and the mine was abandoned.

William Jack and Ed Halliday, calling themselves the Alexandra Gold mining Company, pumped out the shaft, and starting in October 1869 with eight employed men, soon had enough ore at the surface to justify crushing. The very satisfactory return of 2 oz 6 dwt for 1.5 tons of 'stone,'[2] brought about a rush during which nine different parties pegged out claims along the line of the reef. Then there was silence.

This became the common story of the Butchers Gully reefs. Over the next 30 years several other reefs were discovered along the banks of Butchers Creek, and a few were opened up. Someone would find a likely looking seam and would collect a piece of schist rock with gold visible to the naked eye. A claim would be pegged out, a shaft sunk, and that would be the last that was heard of the strike. So it was with the Alexandra reef in the late 1860s and early 1870s, and with the nearby Great Western, the Belle and Beau and the Day Dawn reefs, collectively referred to as the Butchers Gully Reefs.

There are always slow learners when it comes to gold reefs. As late as 1907 a party was giving the Day Dawn reef 'another trial,' and another party in 1910 was using an oil engine to pump out the old shaft of the Alexandra Reef. The results were as before.

CONROYS REEF

Soon after gold was discovered in the gravels of Conroys Creek in October 1862, indications of a reef were seen in the near-vertical walls of the rocky chasm through which the stream flows in the lower part of its course. But such was the abundance of the alluvial gold in this rich gully that little thought was given to quartz mining while wealth could be obtained literally by the shovelful. It was only when the more easily obtained gold became exhausted that attention was finally given to the quartz reef.

It was A. C. Iversen, an already established alluvial miner of Conroys Gully, who first recognised the reef as a potential gold mine. In 1869 he gathered a small group of enthusiasts[3] who were prepared to back the enterprise and sink a number of shafts on the western side of the creek. They obtained reasonable gold but soon they lost the reef, and then water in their deepest shaft caused them to abandon it.

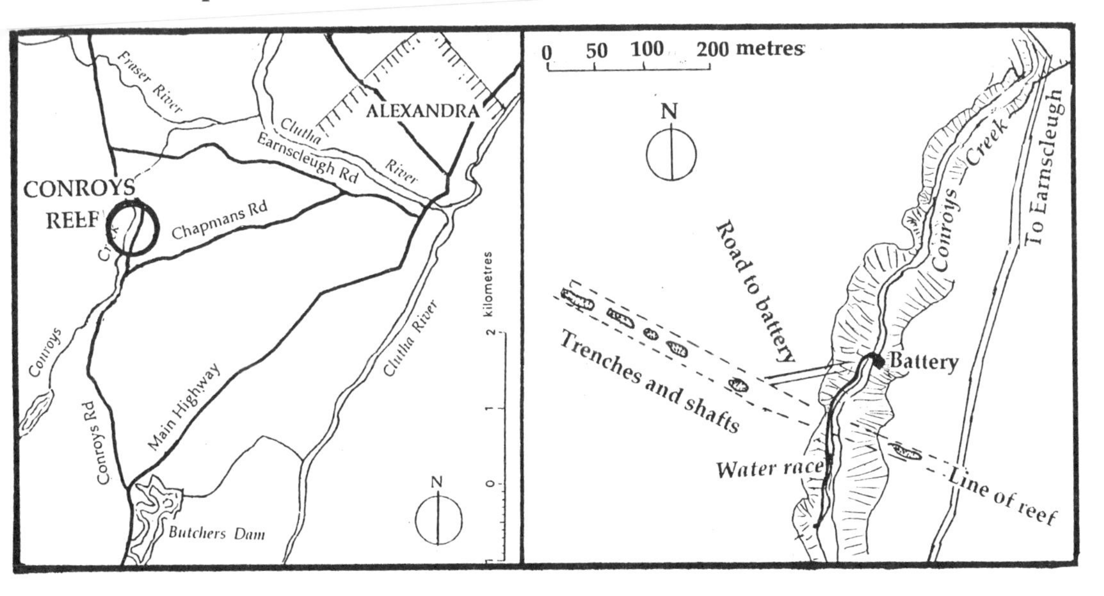

Figure 6.2. Location and principal features of the Conroys Gully reef claim.

In mid-1871 Iverson's party decided on a different approach. Instead of sinking another vertical shaft, they would tunnel in horizontally along the reef. Once they got fairly into the reef the signs were so encouraging that the group decided to set up crushing machinery. A five-stamp battery was erected in the creek bed so that ore, brought from the adit* on a tramway, could be tipped down a shoot directly into the crusher. A substantial fluming of timber and stone conveyed all the water of the creek to the water wheel, which drove the machinery.

The opening of the battery of the Conroy Reef Company was apparently not the occasion that some people had hoped for. A good many invited and uninvited guests turned up, as a rumour had circulated that the christening of the machinery would be accompanied by 'the customary libations'. As the newspaper report put it:

> The rumour proved a mild sell. Those who came eager and athirst for free drinks contented themselves, perforce, by draughts from the cooling creek, and returned, as one disappointed swiper expressed himself, "disgustingly sober."[4]

The drive* was continued for some 600 feet (180 m) and the first

Figure 6. 3. Road (or tramway) to Conroys battery site is arrowed. Conroys Creek, hidden by vegetation, flows in a deep chasm across the foreground.

crushings were most encouraging, yielding 2 oz 13 dwt from each ton of ore. But the problem was to keep up the supply.

By February 1872 the euphoria was obviously dying down with the appearance of reports that Conroys reef was turning out 'to the satisfaction of the shareholders,' which was really damning with faint praise. A month later it was obvious that the writing was on the wall when the best that the newspaper could say was that:

> Conroys reef is not yielding great returns. Six shareholders are working the claim themselves and it is paying expenses.

In November, 1872 the mine closed down. In all, about 500 tons of ore had been crushed yielding £2,500 from the gold obtained, but the expenses, including machinery and labour in driving the long tunnel, amounted to £3,756.

As a consequence the battery, which had cost £850, was sold for £250 and that was the end of the Conroys Quartz Reef Co.

There were plenty of experts to offer advice after the event — notably the Dunstan correspondent of the *Otago Daily Times*, who may well have been a disgruntled shareholder. He was particularly critical of the management of the mine and pointed out that:

> The returns from the top stone were good, and the shareholders thought they were all right. They went to the expense of putting up a nice crushing

plant, but fell into the same error as many companies have done before them. Instead of getting good practical advice, they started their tunnel at too high a level, and as a natural consequence, the top stone was soon worked out; and when it was required to go deeper, they soon found they had water to contend with and that there was not a shaft on the ground capable of being used for bailing water.[5]

The correspondent's opinions were backed by Professor Ulrich of the Otago School of Mines, who pointed out that the drive was only 50 feet (15 m) below the surface when it reached the lode, and consequently there was only a small quantity of ore above that could be worked. The adit should have been 150 feet (45 m) further down the slope. Ulrich also pointed out that a great deal of fine gold, together with a lot of mercury, was lost through lack of understanding of the gold-saving process.[6]

Ulrich, and later Professor Park,[7] also of Otago University, described the reef: it ran east to west across Conroys Creek (which it crossed at right angles) about 220 yards (200 m) downstream from the intersection of Conroys and Chapmans Roads. Traceable on both sides of the creek for a total distance of about 450 yards (400 m), it had clearly defined walls and ranged in width from 6 to 18 inches (15 to 45 cms). Between the walls were more or less parallel thin veins of quartz separated by 'mullock'—crushed schist rock brown in colour and hardened by silica.

Two other Conroys Gully miners, John Bennett and John Dewar, continued to prospect and in April 1879 came across a part of Conroys reef, some distance from the old workings, that had apparently not been examined before.[8] It was a mass of mullock and quartz nearly six feet wide, and there was rising excitement when it was realised that gold could be seen with the naked eye. Nevertheless, in the light of previous experience, the newspaper sounded a note of caution and hoped that any company formed would prospect and test the ground thoroughly before rushing off to buy expensive machinery, as had happened previously.

Bennett, Dewar and four others took out a 16-acre lease over their new find in June 1879, and began driving. They must have been well satisfied with the prospects because, in spite of the well-meant warnings, they began to cast about for ways of financing a crushing plant. In the depressed conditions of the time this proved to be very difficult, and for the first half of 1880 the workings at the reef lay idle while attempts were made to float a company.

Realising that something had to be done to save the enterprise, the redoubtable Jimmy Rivers called a meeting of his friends in mid-June, 1880. The Conroys Gully Quartz Mining Company Ltd was formed with a capital of £6,000,[9] but by the end of July only 1000 shares had been taken up, and by November it was realised that the company was a non-starter.

The claim and gear were put up for auction and bought by Bennett and

Figure 6. 4. The line of the reef is clearly marked by deep trenches and shafts.

Dewar for £41. The pair installed a small battery, driven from an overshot wheel 16 feet in diameter, and quite high gold values were obtained from the few crushings made over the following year. Eventually rising water levels caused abandonment of the drive and the claim, but the battery remained as a reminder of failed projects until 1892 when Crossan and Gray transferred it to their claim on the Old Man Range.

In 1902 John Roberston and J. Paget applied for a lease over the reef and sank a shaft 12 metres deep close to the stream, and then from the bottom of the shaft they drove eastwards along the reef for more than 15 metres in what was described as good quartz.

They were sufficiently confident to buy a 5-stamp battery in mid-1905. Unfortunately, the drive quickly showed that the further in they went the thinner the lode became. Consequently the syndicate stopped work in March 1906 and surrendered the claim later in the year. But apparently this did not entirely prevent Robertson from carrying on mining, as there is a report in 1908 that he was still processing stone when his water supply allowed. Long after he finally ceased working the mine the battery remained and did not finally disappear until after 1920.

OLD MAN RANGE

High up on the slopes of the Old Man Range behind Bald Hill Flat, two prospectors, James White and Andrew Mitchell, found gold in a deposit of soft brown, rubbly schist full of quartz fragments. They applied for a Prospecting Claim in November 1876 and began ground-sluicing the loose debris. By August 1880 they had sluiced a chasm 250 yards (80 m) long and run 500 tons of reef material through their gold-saving sluices. At a depth of 36 feet (11 m), they reached the bottom of the deposit and exposed schist bedrock, but to their surprise they found that the gold-bearing rubble continued down into the rock as a seam about 10 feet (3 m) wide. The two partners had reputedly won £8,000 worth of gold from the reef debris, but George Lythgoe on a claim further down the slope, boasted of having extracted 30 shilling's worth of gold a day from Mitchell and White's tailings.

Professor Park examined the seam several years later[10] and described it as a 'mullock-lode' formed from schist crushed by the movement of a minor fault running in an east to west direction. The southern side of the lode was a near- vertical wall of hard schist that marked the fault. To the

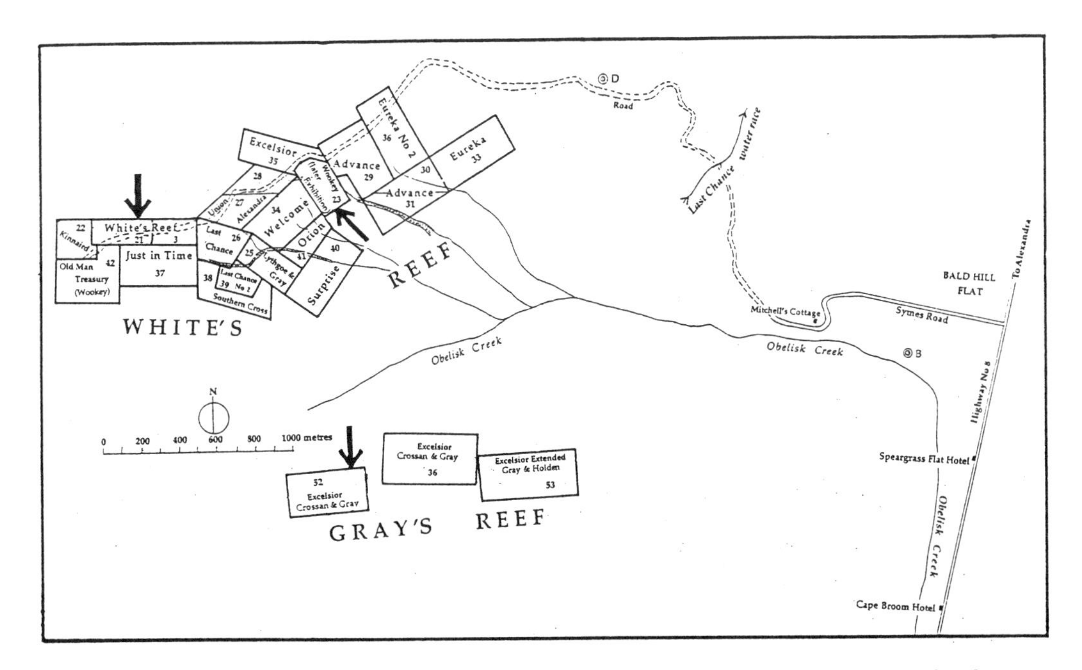

Figure 6.5. Location of claims on the Old Man Range registered during the rush of 1883-84. Arrows point to claims which produced payable quantities of gold.

north the brown, crushed and weathered schist, which could be removed by pick without using explosives, gradually passed into unaltered schist. Park found that the seam of mullock* ranged in width from a few inches up to more than 3 feet. It contained loose gold, which was more plentiful near the surface.

The Rush

The public announcement of White and Mitchell's discovery led to a frenzy of activity during the spring and summer of 1883-84, when no fewer than 20 applications were made for claims.Ü

The competition between parties to get their hands on desirable ground is illustrated by a story related in the newspaper. A party of men from Clyde pegged out a likely claim near White's reef on a Saturday and later in the day another party, already camped on the range, pegged out a portion of the same area. Who would be awarded the claim depended on who was first able to lodge an application at the Warden's Office in Clyde, and then fasten a copy of the application to a peg on the claim.

The local party assumed, correctly, that because the Clyde party lived next door to the Warden's office, they would think that they had plenty of time to put in their application. So the locals made sure that they were on the Warden's doorstep when his office opened on Monday morning.

When the Clyde party got around to making their application they found they had been forestalled. They still had a chance. One jumped on his horse and set off in pursuit of the leader of the local party who was still making his way back to the Old Man Range with his application in his firm grasp. At last he was overtaken and the two men discussed the matter while riding neck and neck. Finally the leader of the local group turned off into the gateway of a friendly farmer as if giving up, only to reappear a few minutes later on a fresh horse. He passed his competitor easily and shortly afterwards fastened the all-important application to a peg on the disputed claim.[11]

As a result of all this activity, it became clear that there were a number of patches of this mullock on the eastern face of the Old Man Range. The greatest concentration was a cluster at an elevation of 1,000 metres on the ridge between Obelisk and Butchers Creeks. The relationship between them is unclear. It has never been properly established whether they were entirely separate entities, or were part of a single structure broken up by earth movements.

Some claims were selected because the presence of rubbly schist on the surface indicated that there might be a buried reef, but a number were taken up simply because the ground was near White's original discovery.

Ü The claims pegged out are listed in an Appendix at the end of the chapter.

Some of these claim-holders had little intention of mining. They were simply waiting in the hope that success in nearby claims would enhance the value of their own properties so they could be sold at a large profit. Others proceeded to sink shafts or commence drives, but it was soon evident that in most claims there was no reef, or if there were, the gold content was far below expectations, or the reef simply 'pinched out.'

There were two great difficulties in mining these reefs. The first was that the reefs were not continuous but were broken up, the result probably of dislocation by landslides. The second difficulty was that the gold was not distributed evenly throughout the mullock but tended to be in patches or 'bunches,' as they were called. Some of these bunches were very rich but were often separated for considerable distances by barren material.

In the end only three claims produced gold in any quantity.

WHITE'S REEF

Shortly after James White and Andrew Mitchell discovered the reef under their claim, White bought Mitchell out for £500, perhaps as a result of a disagreement over the best way to proceed. Mitchell went off and formed a partnership with George Wilkinson in his alluvial claim in Obelisk Creek.

White then made the decision to form a public company to work the mine. A prospectus was published in late 1883[12] and White's Reef Gold-Mining Company Ltd was registered on 13 November. White received 3,000 fully paid up £1 shares for his mine and the remaining 7,000 shares were quickly taken up. Of the 61 shareholders, more than half

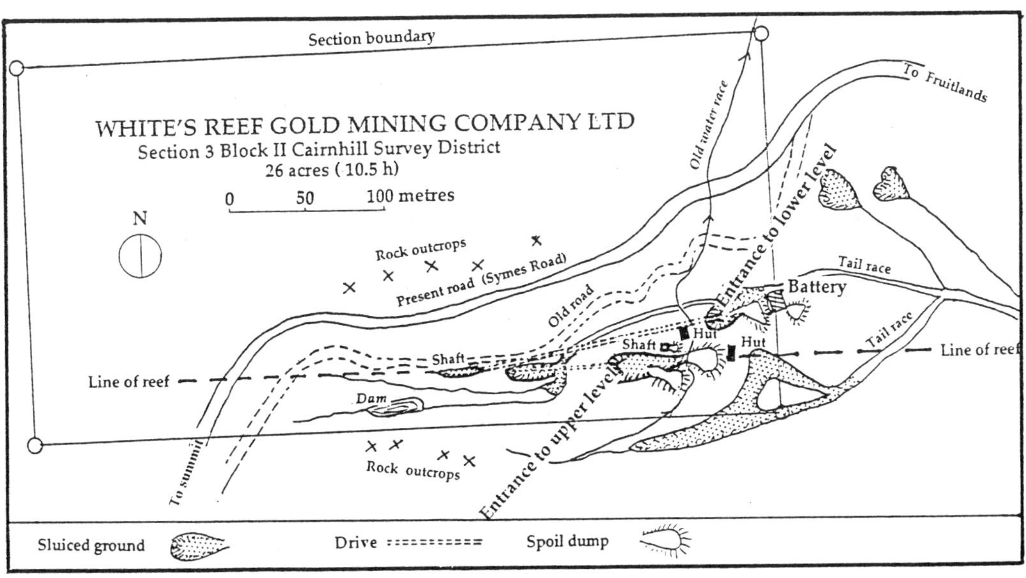

Figure 6. 6. Plan of White's reef drawn from air photographs, field sketches and descriptions.

were from Alexandra or Bald Hill Flat but the directors were mainly Dunedin businessmen. J. B. Neal was appointed manager.

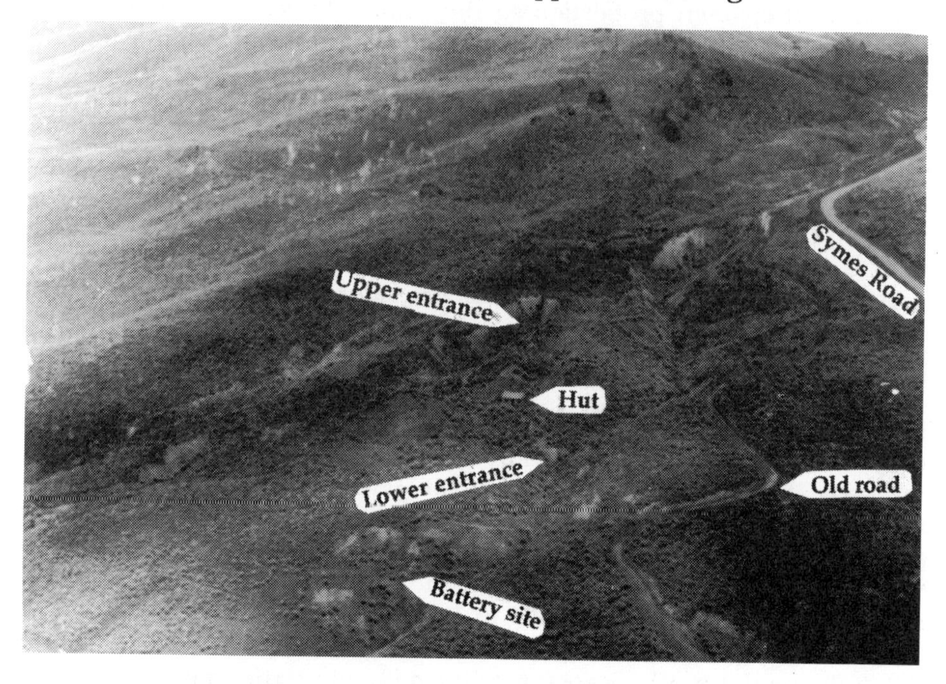

Figure 6. 7. White's reef workings from the air.

A blunder was made at the very beginning. The position selected for the adit was at too shallow a depth so that there was only a small amount of gold-bearing material between the drive and the surface. Even though he had received instructions from Dunedin as to where exactly to place the adit, Neal, the manager, was blamed for the mistake and sent on his way. Thomas Andrew, of the neighbouring Advance Company, was the next manager and a man with a gift for writing optimistic reports. He asked repeatedly for a stamper-battery to crush the ore but the directors refused to consider one until 500 tons of ore were 'at grass.' It was not long before just such a quantity had been accumulated at the mouth of the tunnel and the directors were reminded of the battery deal.

The 5-head stamper battery from the Lucknow mine at Bendigo in the Upper Clutha Valley was bought and erected by February 1886. It was to be driven by a Whitelaw water turbine fed by water races from Obelisk and Butchers Creeks, but they proved inadequate and the battery could only operate for about six hours each day — if it did operate at all.

Shareholders were greatly looking forward to the results of crushing the stockpile,as they had been led to believe that it contained large quantities of gold that would clear all debts and pay a handsome dividend. So it was particularly frustrating when the battery suffered one breakage after another. There was at least one holdup every week until the middle of the year. And after all this the results of the crushing were very poor. It seemed that the 500 tons of so-called 'ore' contained a great deal of stone that was short on gold.

It was reported after this crushing that Mr Andrews was encouraged to seek fresh fields where there were no batteries, and most of the men were sent with him. A temporary manager, J. Mitchell, succeeded in finding some rich ore, but after a difference of opinion with the directors he resigned after a few months. As the newspaper remarked, 'Management from a distance does not work.' Taking advantage of Mitchell's find, the next manager, J. Bennett of Conroys Gully, appointed in early 1887, managed an excellent return of 580 oz of gold from 600 tons of ore. It cleared all debt and the company was even able to pay a small dividend. Then in late 1886, two men who were to greatly influence the future of the company were appointed to the staff, the brothers Robert and Henry Symes.

For the next four years the White's Reef Company prospered. Employing 20 men and boys the mine was producing gold at the rate of 1,000 oz each year. Ore was averaging 1 1/2 oz of gold per ton and the mine was paying dividends. Shares on which 10 shillings had been paid up and which previously could not be given away were selling at 12 shillings. It was during this period that White's reef gained its reputation as one of Otago's foremost gold mines. There were very few other quartz reef mines that were consistently returning over one ounce to the ton of stone crushed. Nevertheless dividends were small.

Professor Park examined the mine in November 1888 and wrote a scathing report.[13] At this stage the main drive was 740 feet (225 m) long with three vertical ventilation shafts spaced along its length. Because of the shortage of material overhead, the company was forced to mine from below the floor of the drive — a most difficult task as material had to be hoisted up a 40 foot (12 m) shaft, and because of lack of drainage, miners had to work in very wet conditions. Park listed the main reasons for the poor returns to shareholders as the short working year owing to closing the mine during the winter, and the heavy maintenance of the long drive[14] which was a serious drain on the resources of the company. To these he added the narrowness of the lode, and the shattered and dangerous nature of the country which required the use of much heavy, high-priced timber.

At the beginning of 1889, Robert Symes was appointed manager in place of Bennett, who had resigned through ill health. Then suddenly the gold ran out.

The directors ordered the men's wages reduced from 10 shillings a day to 8 shillings, so they went on strike. There were plenty of others to take

Figure 6. 8 This sluiced gully is probably the site where White and Mitchell discovered White's reef. The rock wall on the left side is a fault scarp. Schist debris in the foreground is from a shaft.

their places but the newcomers also demanded 10 shillings a day and got it.

It was decided by the directors to extend the drive a further 200 feet (60 m) at a cost of £250, in the hope of finding gold again. Hopes were raised when a rich patch was encountered, but it was only a pocket yielding £95 worth of gold. Manager Symes advised against spending any more money, so all hands were discharged, and as the losses already amounted to £2,500, the mine closed in September 1890. The following month, the mine and other assets were sold by auction amid questions asked by shareholders about the legality of the sale, as they had not been consulted by the directors beforehand.[15] Robert and Henry Symes bought the mine and with it a five-stamp battery and other machinery, mining tools, buildings and water rights, all for £250.

Symes' Mine.

The Symes brothers set about remedying many of the deficiencies in management that had plagued this mine, rich in gold but poor in dividends. New gold-saving machinery, a more efficient turbine, timber obtained from Waikaia Bush at half the price of that from Tapanui, and a system of re-using mine timber that was a great financial saving, were some of the improvements. Robert Symes was mine manager and brother

Figure 6. 9. Symes' battery at White's reef.
Upper: Battery site. The vertical shaft (centre) is the Whitelaw turbine. The Excelsior reef is on the ridge in the background.
Lower: The Whitelaw turbine runner (centre) and its base (right).

Henry managed the stamper-battery which included the gold-saving devices.

They were not able at this stage to open up new ground to search for the missing reef, so concentrated on cutting out pockets of good ore that had been overlooked. In this way they recovered £1,000 worth of gold over the next 12 months. To strengthen the partnership financially, they sold one third of their holdings to Thomas Andrew of Tasmania, the former manager, for £280.

In 1893 Symes was able to sink a shaft about 250 feet (80 m) ahead of the old workings and was fortunate to strike the reef again at a depth of 30 feet (9 m). The original drive was then extended to cut this shaft, so ore could be run out to the battery. This year 130 tons of ore yielded 119 oz of gold.

It was unfortunate that in 1896 a workman, C. E. Jones, was killed while stripping surface material by sluicing. He was working alone and details are scarce but it is assumed that he was struck by falling debris.

Meanwhile the reef continued to be worked by sinking shafts and then driving between them. In 1897, perhaps the last good year, 200 tons of stone were crushed to give 270 oz of gold.

The closeness of the main drive to the surface, which forced Symes to extract ore from deep below the floor, was a great disadvantage to the working of the mine.

Finally a decision was made to excavate a small, prospecting drive from about 130 feet (40 m) further down the hill. In anticipation of the arrival of much gold-bearing stone, the stamper-battery was, in 1898, shifted down to the proposed mouth of the new tunnel. The extra fall also increased the power from the turbine. In 1899 six men began work on the new adit.

This new drive had to pass through a considerable distance of solid 'dead' rock before the lode could be reached, but by 1901 it had been driven 400 feet (120 m) and had reached the reef. It was then continued

Figure 6. 10. Robert Symes.

along the lode, and a year later was in 650 feet (200 m) but still no substantial body of gold-bearing stone had been encountered. There was a delay while a necessary ventilation shaft was cut up to the original drive 70 feet above but by 1903 the new drive was in 750 feet (230 m). However, payable gold was sparse. Symes kept coming across small 'stringers' of gold-bearing quartz that had the effect of leading him on and encouraging him to continue with this hard, time-consuming and dangerous work.

Symes finally abandoned underground working at White's reef in 1906. But this did not mean that he had finished with the mine. He went back to ground sluicing, working his way along White's reef, and then began work on the Advance claim half a mile (720 m) east of White's reef. It was not long before he unearthed a patch of good stone in a corner of this claim and could not resist the temptation to put in a short drive. He bought the small battery from Gray's claim, which had closed down, and set it up in a gully on the Advance claim.

For the next 20 years, with two employed men, Symes continued sluicing wherever he thought there might be gold bearing ore, saving any he uncovered for crushing in his little stamper battery. He finally gave up mining in 1927.

White's Reef today

The most conspicuous feature of the workings at White's Reef today is a well-constructed stone hut that stands about 100 metres to the south of Symes Road. The present corrugated iron roof was put on by the newly formed Vincent Ski Club during the 1950s. Uphill from the hut are two deep, narrow gashes in the hillside. The upper one, about 150 metres above the hut, may well have been the site where White and Mitchell first discovered gold. It appears to have been formed by sluicing only, as the absence of any spoil dumps indicates that there was no drive from this chasm. The lower excavation, closer to the hut, does have spoil dumps associated with it, and was almost certainly the site of the White's Reef Gold Mining Company's drive. The original site of the stamper-battery was close by.

In the gully which lies between the hut and Symes Road and less than 100 metres down the hill from the hut, is a sluiced excavation and a spoil heap which mark the site of the entrance to the lower level drive. Just below the spoil heap is a flattened platform which marks the site of Symes' main battery. Lying about are the remains of machinery, including the runner of a Whitelaw turbine.

On all sides of the workings, sluiced excavations mark places where prospectors, including perhaps Robert Symes in his later years, attempted to find the continuation of known reefs or to discover new ones.

Over 700 metres in a north-easterly direction from White's Reef hut, in another gully, lies debris of mining activity including the remains of a small building which may have housed a Pelton wheel, a sluiced

Figure 6. 11. Remains of two stamper-batteries on the Old Man Range.
Upper: Small structure that probably housed the Pelton wheel which drove Syme's last battery.
Lower: 3-head stamper camshaft and stamper shoes probably from Exhibition battery.

excavation and a small spoil heap. This probably marks the site of Robert Symes' small mine on the Advance claim.

EXHIBITION REEF

In December 1882, William Wookey and James Gavin, encouraged by White's success, began sluicing an outcrop of rubbly schist with loose gold 650 yards (600m) north-east of White's claim. They had the experience of James White to guide them, so they set up a dam to catch their gold-rich quartz tailings. It was known by this time that White had lost much gold in the small fragments of quartz he had discarded. Wookey and Gavin were sufficiently convinced of the value of their ground to apply for a 10-acre lease in July 1883.

The mine underwent a series of changes of ownership as partners bought in and dropped out, but some stability was reached in 1887 under Hugh Crossan and James Gavin. Using the battery at White's Reef, they managed to recover 2 oz of gold from each ton of ore. It was at this time that the claim was named 'Exhibition Reef.' It was at this time also that one of the men, swirling some of the crushed ore from the mine around in his hat, suddenly collapsed. It was surmised that he was overcome from fumes from the arsenopyrite, one of the minerals occurring with the gold.

Crossan went bankrupt in February 1891 and the claim was bought by Dr Hyde and Henry Symes. Finally Henry Symes became the sole owner. A small 3-head stamper, driven by a small wooden water wheel, was erected in September 1893 and found itself busy day and night when a shaft, sunk to 30 feet, came across a very rich patch. In 1895 the mine closed down pending the purchase of the 10-head battery erected 12 years before in the head of the Fraser River by the Alpine Quartz Company. It was never installed in the Exhibition mine, which did not reopen. The claim was abandoned in 1898.

Exhibition Claim today.

About 600 metres north-east of White's reef and about 100 metres up the gully from Symes battery site on the Advance claim, are the remains of a 3-head stamper-battery. A camshaft with its three cams, a few stamper shoes, sheets of iron and some timber are scattered about on low stone walls. A sluiced excavation and a spoil heap indicate that here was once a drive or shaft nearby. This is almost certainly the remains of the Exhibition battery .

EXCELSIOR MINE (Grays Reef)

Hugh Crossan and Frank Gray discovered a reef, which over its short life produced more gold than any other quartz mine in the district. It was on the ridge between Coal Creek and Obelisk Creek about a mile due south of White's Reef, and 200 metres lower in altitude. They applied for the claim in January 1891.

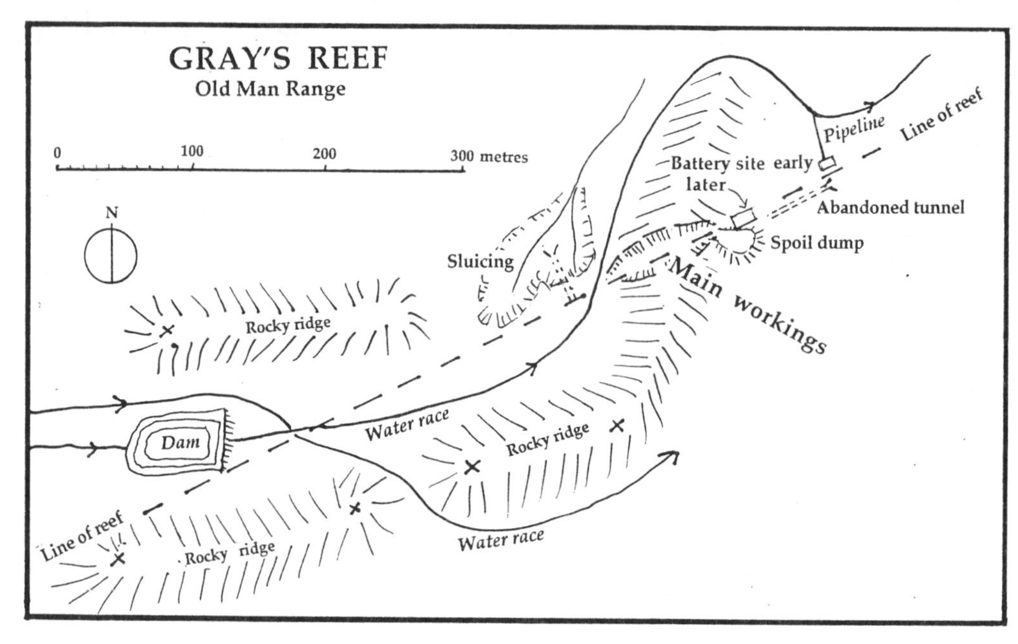

Figure 6. 12. Plan of Excelsior mine drawn from air photographs, field sketches and descriptions.

Using Kemp's water race for sluicing, they exposed, at a depth of 16 feet and over a length of 100 yards (90 m), a lode of friable quartz in soft schist, confined between hard schist walls 3 to 10 feet apart. They sank a shaft 20 feet (6 m) into this lode and began to lift ore by means of a hand windlass. In six months they recovered 766 oz of gold from 315 tons of mullock. Including the 114 oz from the sluicing, this gave a total of 880 oz recovered for the year.

As a result of tests carried out after sinking a shaft 65 feet (20 m) deep, the partners went over to Conroys Gully and brought back the small 3-head stamper battery that had been set up by the Conroys Quartz Mining Company in 1880. Water piped from a water race 100 feet above, produced 3 horsepower from a Pelton wheel about three feet in diameter. This battery could crush up to 3 tons of ore in an 8-hour day.

Now that the gold-bearing quartz from the lode could be crushed, the return of gold increased remarkably. A return of 256 oz from 118 tons of rock crushed (more than 2 oz to the ton) was reported in December 1892, and from mid-October 1892 to mid May 1893, 918 oz of retorted gold worth over £3,700 was obtained. Little wonder this mine was described by an enthusiastic correspondent as 'the most valuable mining property in New Zealand.'[16] Work stopped for the months June, July and August when the water sources were frozen.

The mine in 1893 was still being worked by primitive methods and with a minimum of labour. Shafts up to 40 feet deep were sunk at intervals along the reef with drives connecting them. Only one man worked in the drive breaking out ore which was lifted to the surface by hand windlass. It was then sledded to the battery that Gray had located some distance down the hill from the mine in anticipation of starting a low-level drive. In late 1893 work on this began. The intention was to drive, from the steep slope below the mine near the battery, along the reef at a depth of 100 feet (30 m) below the surface workings. Some 250 feet (75 m) in, however, the ground became unstable, and earth movements caused the heavy mine timbers to shatter so work was stopped. Nevertheless after a slow start, 1896 turned out to be a good year with 270 oz of gold produced by two men over a few months. More than 2,000 oz of gold had been produced since the mine opened. The low-level drive was abandoned in 1898 and a disappointed Hugh Crossan sold his share of the mine to his partner for £250 and went off to buy the Beaumont Hotel.

Figure 6. 13. Sluicing the reef material left this large trench, about 80 metres long and 6 metres deep. It cuts right through the ridge and is the most conspicuous feature of the Excelsior mine today.

It is interesting to note that during all the time Frank Gray was working at this claim, he continued to live high up on the hillside above Bald Hill Flat ('above the fog' he was reputed to have said), so each morning he had not only to walk two and a half miles (4 km) but also descend 100 metres into Obelisk Creek and then climb 300 metres up to his claim.

After the low-level drive was abandoned, the battery was moved up the hill to a more convenient site, and Frank Gray went back to sinking shafts and lifting the material laboriously by hand. One disadvantage of working down from the surface was that all the rainwater and melting snow ran into the shafts.

In 1903 another attempt was made with four employed men to drive into the reef. This time a 'cross-drive,' as it was called, was driven at right angles to the reef from the steep northern slope of the ridge. After cutting through 300 feet of broken schist the men finally reached the reef and then proceeded to drive along it. Some good gold was obtained — one of the best returns was 62 oz of gold for 28 tons crushed but the lode then pinched out. Francis Gray sold the little battery to Robert Symes and abandoned the mine. £9,800 worth of gold had been recovered for the seven seasons worked and when Gray retired in 1907, he was a wealthy man.

Excelsior Claim today

The most noticeable feature of the old Excelsior claim is a large gash in the ridge. This gash looks not unlike the cutting made for a railway, but

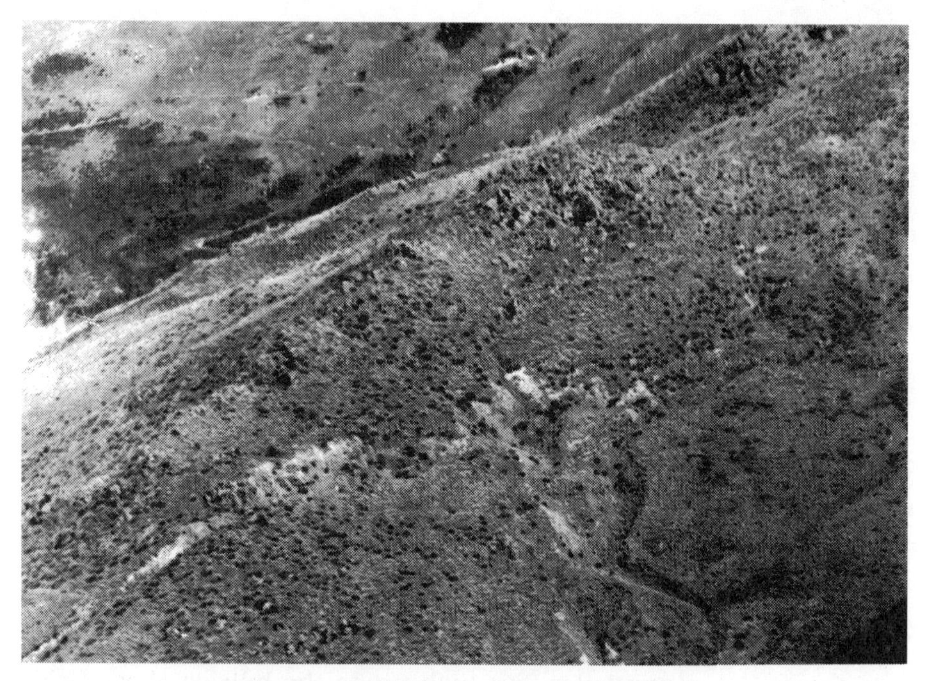

Figure 6. 14. Air view of Excelsior mine.

was formed by Gray and Crossan as they sluiced away the soft material of the reef. A dump of spoil at the eastern end of the cut was derived from sinking shafts in the floor of the cutting and from the drives running away from the bottom of the shafts. Below the dump is a flat platform that was the latest site for the stamper-battery. An earlier battery site was 100 metres down the steep hillside near another large dump of spoil derived from the various attempts to excavate a low-level drive along the reef. Sluicing on the north side of the ridge conceals the entrance to the cross-drive. Several hundred metres west is the dam that supplied the water for driving the pelton wheel of the battery and for sluicing.

NOTES

1. The site of the main shaft is shown on some old mining maps. It was less than 200 metres from the Alexandra – Butchers Creek road on the western side. Other prospecting shafts overlooking the gorge of Butchers Creek are still visible.

2. The material containing the gold and set aside for crushing, was always referred to as 'stone' whether it was hard quartz or soft decomposed schist.

3. The members of the party were: Iversen, C. Iversen, W. Grindly, Stephen Foxwell, William Beattie, and Thomas Oliver.

4. *Otago Daily Times* 30 August, 1871.

5. *Otago Daily Times* 6 November, 1872.

6. Ulrich Votes & Proceedings Vol XXXIV 1875 Appendix pp. 76-77.

7. J. Park 1906 pp. 30-31.

8. *Dunstan Times* 25 April and 22 August 1879.

9. *Dunstan Times* 20 April 1880.

10. J. Park 1890 pp. 32-33.

11. *Dunstan Times* 15 February 1884,

12. *Otago Daily Times* 12 October 1883.

13. J. Park 1890 pp. 32-33.

14. A report by the Inspector of Mines commented: "The upper level was constructed on such a serpentine course that it seemed to any one acquainted with mining that those in charge of the work at that time had no idea where the lode was to be found." AJHR 1895 C—3 p. 87.

15. *Dunstan Times* 7 November 1890.

16. *Dunstan Times* 20 January 1893.

APPENDIX

Applications for reef claims on the slopes of the Old Man Range above Bald Hill Flat are listed in alphabetical order with dates when they were first granted and principal applicants. On most of these claims prospecting showed no signs of a reef and the claims were never worked:

Advance Quartz Mining Co: October 1883

James White, Geo Wilkinson, Thomas Rahill, John R. Kemp, William Fraser, Ewen Pilling, James Gavin, Robt Kinnaird

Worked again by Robert Symes in the 1920s up to 1927 which is the last record of any work on the Old Man Range reefs.

Alexandra Gold Mining Co: December 1888

Took over the Union claim below Whites Reef. By March 1889 a tunnel was in 310 ft and they were prepared to go another 100 ft but there are no

further records.
Eureka Gold & Quartz Mining Co December 1883
Richard Melner, R. J. Pitchers, J. Steele, et al.
Eureka No 2 : December 1883
James McCormick, Charles Bayford,William Wookey, Charles Gibson and two others
Excelsior Co: December 1883
James McCormick, Charles Bayford, William Wookey, Charles Gibson and two others
Excelsior : July 1883
H. Crossan, F. W. Gray
Exhibition : June 1883
William Wookey, James Gavin
Great Western Co: February 1894
James McCormick, Charles Rayford, Charles Gibson, John Butler
Just in Time Co: February 1884
W Quayle, John Ryan, C. Cooper, W. C Hodges
Last Chance Co: October 1883
Bernard O'Neill
Lythgoe & Gray: October 1883
George Lythgoe, James Gordon Gray
Kinnaird & Co: June 1883
Robt Kinnaird, John Dewar, John Bennett
Old Man Range Co: September 1884
Butler, Gibson, McLusky, H. Crossan, Geo Lythgoe, James Hesson, W. Lynch
Old Man Treasury Gold Mining Co: September 1885
W. Wookey
Orion Gold Mining Co: February 1883
John Baker, Charles Holden, John Dickie
Southern Cross: February 1884
Surprise Co: February 1884
Geo Lythgoe
Union Gold Mining Co: November 1883
Frances Baker, Blackwell
Welcome Quartz Mining Co: November 1893
15 shareholders including John Butler, Charles Gibson, James McLlusky (McCloskey?), Hugh Crossan, Geo Lythgoe, William Lynch, James Hesson

7.

WAR IN THE GULLY

Coal at Clyde—and elsewhere

Miners rushing to the new gold discovery at Hartleys Beach could not fail to notice the prominent seam of coal clearly visible along the bank of the Clutha River at the mouth of the Dunstan Gorge. It was not long before miners were helping themselves to the free fuel that was much appreciated in the treeless countryside in which they found themselves. The tent town that was being established on the terrace above was for a time referred to as 'Coal Point,' but then underwent several quick name changes until finally saddled with 'Clyde.'

The seam of coal, some nine metres thick, was clearly exposed for about 250 yards (75 m) along the eastern bank of the Clutha River above the present Clyde bridge. The Provincial Government, with the coming winter in mind, quickly called for tenders[1] for the lease of the coal outcrop. Conditions were that 100 tons of coal was to be supplied each month at the pit head at 20 shillings a ton.

Dunstan Coal Pit

J. Morgan and party took up the lease and began to mine by opencast methods. But the mine was difficult to work as it was close to the river, and was inundated by each flood. In fact after the big flood of 1863 had left the people of the Dunstan without firing in the middle of one of the worst winters on record, there was strong criticism of the mine management. As the easily worked coal was already running out, Morgan gave up and sold the mine to James Holt in May 1864.

James Holt was already a prominent figure at the Dunstan. He had been a member of the first party to leave Gabriels Gully for the new goldfield in the Rush and has earned his place in Dunstan folk-lore as the man who assisted Mrs Sarah Cameron with her young children on that memorable journey over the mountains. After a stint of gold mining on the river beaches, Holt, with his friend Roscoe, spent the first half of 1863 constructing a water race from the gorge of Waikerikeri Creek to Clyde

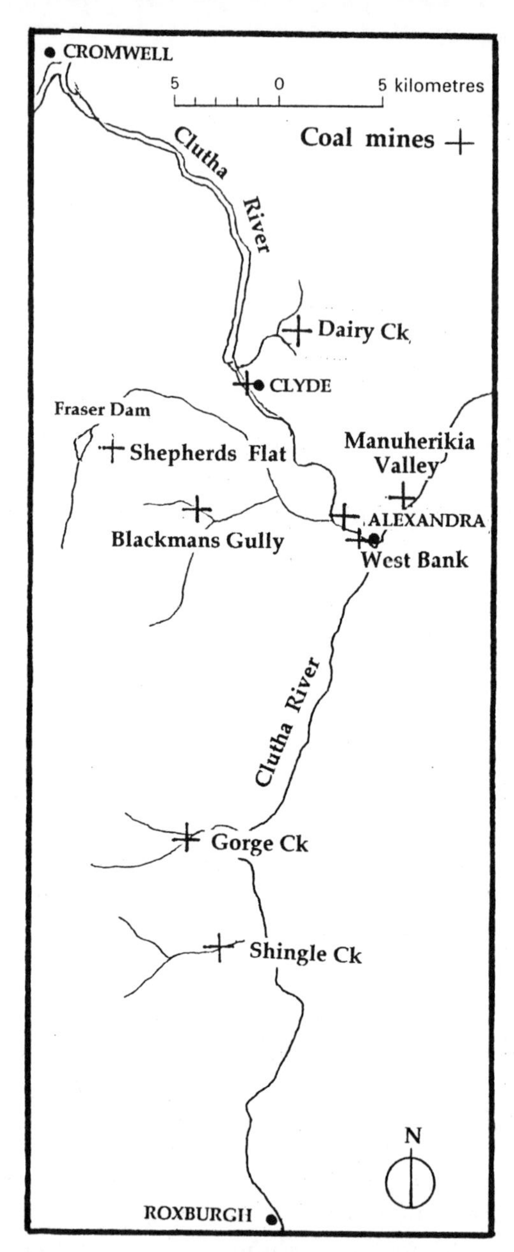

Figure 7. 1. Location of coal mines in the Clyde - Alexandra district. Coal mines near Alexandra have been described previously (McCraw, 2000).

township, where the water supplied the domestic wants of the town and turned a small water wheel to work the pumps at Morgan's Dunstan Coal Pit.

When Holt and partner, Dakin, took over the Dunstan Coal pit they lost no time in sinking a shaft and installing efficient pumps. Within a couple

Figure 7. 2. This 1865 photograph of Clyde shows James Holt's coal mine on the banks of the Clutha River. Various features are marked including the water wheel which drove the pumps.

of months they were supplying coal from 'a splendid seam' at a depth of 32 feet (10 m).

Their system of operating the necessary pumps was unusual. Water from the race, which Holt owned, first passed over a small water wheel which operated a Californian pump.* It then passed on to the main three metre-diameter wheel, sited at a lower level, which worked a barrel pump capable of clearing the mine in six hours. The Californian pump lifted some of the water that had passed over the large wheel back into the head race where it was recycled, and was said to increase the power of the big wheel by one half. The whole scheme cost Holt £1,100, of which the water race accounted for £600 and the machinery the remainder.

The partners had their problems. Twice they had sunk shafts and twice they had collapsed. The third was successful, but they still had difficulties with water. During a particularly high flood, Holt was forced to turn off the pumps and allow the mine to fill with water, rather than risk the serious damage that could be caused by the floodwaters overtopping the shaft and rushing into the mine. Then there was near disaster in late 1875 when the river broke into the workings, causing the miners to flee for their lives.

These difficulties, coupled with the constant and increasing harassment by farmers wanting to use the water from Waikerikeri Creek (which supplied his water race) for irrigation, were wearing Holt down. He had already decided to shift his main operations to an easier place before the great flood of 1878 inundated the mine.[2]

Dairy Creek.

It was in Coopers Gully, a small tributary of Dairy Creek[3] the first creek to enter the Clutha River upstream from Clyde, that coal was discovered during ground sluicing for gold. It was to Coopers Gully that James Holt decided to transfer his coal mining operations in 1880. The site had the decided advantage that it was at least 90 metres above the river, but Holt did not have the gully to himself. It was already occupied by a coal miner, Collins Toussaint Marie.[4]

Collins Toussaint Marie and his friend, La Fontaine, had first attracted attention at Clyde when it was reported[5] that 'two enterprising Frenchmen' had taken up a 4 acre claim on the river bank below the Police camp. They had ordered, from Dunedin, the necessary material to construct a 'water-lifting machine on the Thomson principle'[6] to raise water from the river to their claim. This was the last we hear of Marie's claim and his water lifting machine, but it was not the last we hear of Marie.

Collins Marie had become interested in coal mining, and apparently tried to gain access to Holt's mine on the river bank at Clyde, presumably to see what Holt was up to. It is likely that he was ejected from the property because his lawyers had then approached the Waste Lands Board for authority for Marie to enter and view Holt's mine. This, the Waste Lands Board was not prepared to give, but instead, asked that a Government officer inspect the mine at Marie's expense, and report to the Board. At the meeting of the Waste Lands Board in August 1871 Marie was granted a coal lease over 5 acres in Dairy Creek. The price of coal was set at 15s. a ton at the pit mouth.[7]

War in the Gully

Marie constructed a road, which he regarded as his property, up Dairy Creek and opened up his 'Dairy Creek' mine in Coopers Gully. Eight years later when Holt first applied for a coal lease in Coopers Gully, a roadline, more or less following Marie's road, was surveyed up the gully. To provide access to Holt's lease further up Coopers Gully, the surveyors laid out a branch roadline along the boundary of Marie's lease. The road that Holt formed along this roadline to his mine, although not on Marie's lease, cut across the entrance to Marie's mine. Marie promptly put a fence across Holt's road. So began a conflict between the two miners that was reminiscent of the vendettas in the Appalachian Mountains of Virginia.[8]

Holt complained to the Vincent County Council and, with the tacit support of the council officials, he pulled down the fence and continued to extend his road, damaging Marie's orchard in the process. In an

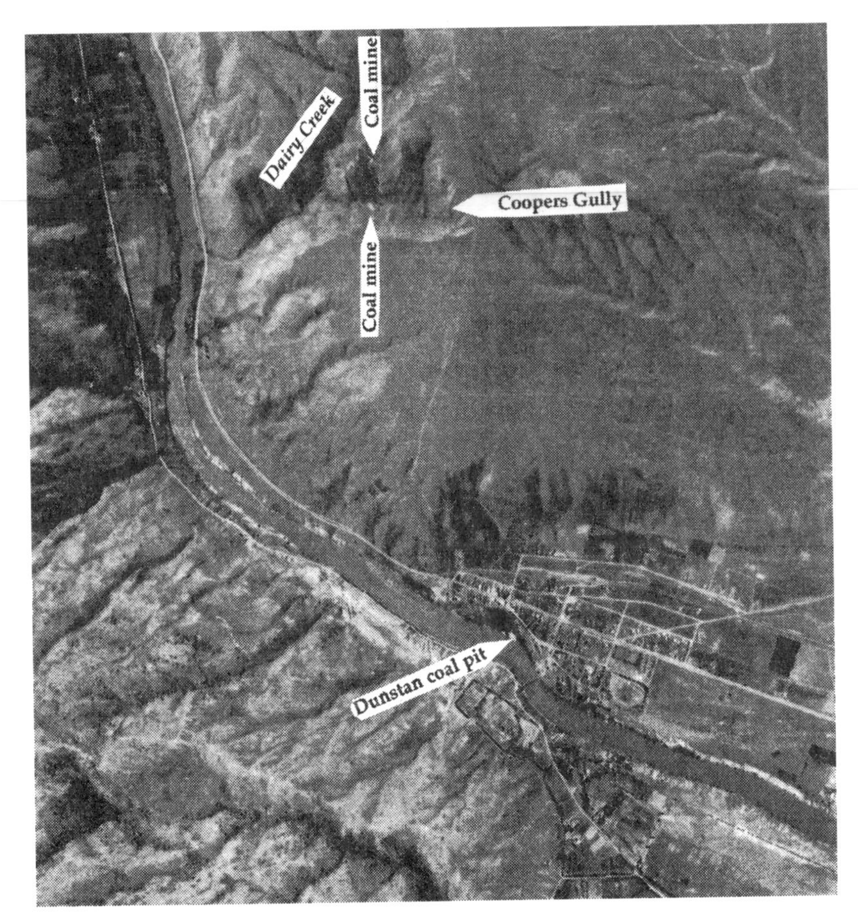

Figure 7.3. A 1958 air photograph shows Dairy Creek and Coopers Gully in relation to Clyde and the Clutha River. The site of the Dunstan coal pit is marked.

attempt to solve the problem, the County Council surveyed a new roadline to Holt's 'Waikerikeri' mine in 1882 which was well clear of Marie's mine, but as Holt had already built his road he was in no mood to build another.

Marie blocked the road again and took the council to court. The judge gave the council officers a ticking off over the way they had treated a man who had lived in the gully for 15 years and had built a road in to it at his own expense.

Eventually, in 1887, the road up Dairy Creek to the mines was made a public road, but Marie challenged this decision in court and was able to hold things up. He actually wrote to the Inspector of Coal Mines asking him to interview the Minister of Mines about the road that had been constructed in front of Marie's mine. The inspector replied that his duties

Figure 7. 4. The road up Dairy Creek, first constructed by C. T. Marie, is still clearly Visible but its lower end is now submerged by Lake Dunstan.

did not include interference in disputes of this nature. Two years later Marie wrote to the inspector again telling him that Holt had allowed a fire to spread into Marie's mine and asking that the inspector inquire into the matter. Again the inspector replied that it was impossible for him to interfere and pointed out the relevant section of the Act under which Marie could take action if he wished.

In 1891 Marie and 125 others petitioned Parliament asking that the road to Holt's mine be closed and that Marie be paid compensation for building the road from Clyde. The Public Petitions M-Z Committee recommended on 21 September that the petition be referred to the Government for consideration, but nothing happened. A second petition a year later met with the same fate.[9]

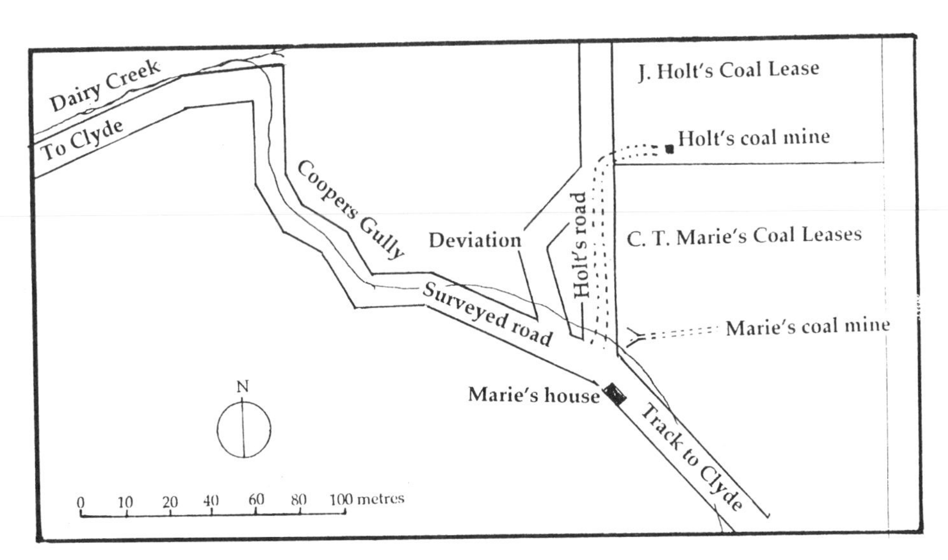

Figure 7. 5. Coal mines and roads in Coopers Gully in 1887. The road has been surveyed from Clyde and extended to Holt's mine in the upper gully. Holt's road interfered with Marie's mine, so a deviation was surveyed.

Dairy Creek mine

Collins Marie's Dairy Creek mine was on the floor of Coopers Gully and was entered by an inclined adit. After a distance of only 30 feet (9 m), it reached the coal seam which was 20 feet (6 m) thick and dipping gently to the southeast. Marie worked his mine on the 'bord and pillar' system, which meant he extracted coal from a pattern of corridors at right angles to each other, leaving substantial pillars of coal between the corridors as support for the roof. A plan of the mine looked similar to that of a city with its blocks and streets. Coal was loaded into trolleys that were drawn out of the mine by a hand-turned winch. The output from the mine, when everything was going well, was between one and two hundred tons a year.

Marie's mine was subject to spontaneous combustion, and when fire broke out the mine had to be boarded up until it died down. Apparently he didn't know much about organising a coalmine, and was the despair of the Inspector of Coal Mines. In 1881 the inspector wrote:

> This mine, belonging to Mr Collins, (sic) is still worked as badly as ever. The Act is not observed in any way, but the report-book is said to have been burnt in a recent fire. And this cause is also assigned for the neglect to exhibit or distribute Special and General Rules.[10]

Because of the fire, Marie opened a new drive in 1884. It began to heat also, but he was still able to take out up to 200 tons of coal a year. Four years later the inspector was still complaining that Marie had done nothing towards complying with the Act.

None of these problems curtailed Marie's ambitions, however. He applied for a 17acre block between Coopers Gully and Dairy Creek under a section of the mining Act which allowed miners who had demonstrated that they had 'settled down' by establishing a 'homestead,' to purchase up to 50 acres. A 'homestead' had to have not only a dwelling, but also some evidence of agricultural activity such as a dairy, market garden or orchard. The real reason for the application is clearly shown on the accompanying plan — there is drawn a 'proposed tunnel,' nearly 300 metres long, from the valley of Dairy Creek, through a ridge of solid schist, to Marie's coal mining lease in Coopers Gully.

Marie did not go ahead with this drive which would have enabled him to tap the coal seam at a considerable depth. However, a drive following nearly the same route, and of the same length, was driven a few years later by the Dunstan Coal Company in an endeavour to reach deep coal in Holt's lease.

Figure 7.6. The ruins of a stone building marked on Park's 1906 map (Figure 7.8) as 'Mine house,' and on old mining maps as 'Marie's dwelling.'

Because of persistent fires, Marie's mine was virtually closed down from 1887, but every so often he would open it up and attempt to extract some coal. Inevitably the fire would flare up again. The inspector wanted him to close the mine and start a new one at a distance from the burning coal, but Marie did nothing. In the end the inspector recommended to the

Commissioner of Crown Lands that Marie's lease be cancelled, but then G. Griffiths, who had taken over the neighbouring Holt's mine, offered to help Marie straighten up his mine. The inspector was pleased and asked the Commissioner to stay the Cancellation-of-Lease order. The inspector's hopes were premature. Only a small amount of work was done before Marie and Griffiths parted company. The frustrated inspector wrote:

I do not now think the pit will ever be properly worked by Marie. I can do nothing more to help him. I have lost faith.[11]

Eventually, J. Pratt entered into some kind of partnership with Marie and took over management of the mine in the late 1890s, but he was able to extract only 678 tons of coal before he too was on his way. It was at this stage that Collins Marie, after nearly 30 years, finally abandoned mining.

W. J. Tonkin took up the lease in early 1900, and with G. Robertson as manager, decided to abandon Marie's mine and open up a new one a short distance away. Robertson sank two shafts 130 feet (40 m) deep and connected them underground by drives on two levels. Then he constructed an incline from the surface to the lower drive. But the greatest innovation was the introduction of a steam engine to haul the coal trolleys from the mines, instead of using horses or the laborious hand winches.

Coal production was raised to over 1,000 tons a year, rising to over 2,000 tons in 1902, to meet the demand for coal from the increasing number of gold dredges. In March 1903 the Dairy Creek mine was taken over by the Clyde Collieries Company, but continued for a short time as a separate entity.

Figure 7. 7. The Dairy Creek mine, under the management of R. Robertson, replaced horse or manpower haulage by a steam engine in 1902.

Waikerikeri Mine

In spite of the almost open warfare between Holt and Marie, the winning of coal from Holt's mine was uninterrupted. Holt was working in a 20 foot (6 m) thick seam which stood at an angle of 45^{0} dipping to the south. Coal at the rate of 300 to 400 tons a year was lifted up a 100-foot (30 m)-deep shaft by a horse-driven winch.

On a Saturday afternoon in March 1891 James Holt and an employee, John Statham,[12] were working at the foot of the shaft when they were both overcome by foul air. Holt passed out but Statham managed to grasp the rope of the windlass and was hauled to the surface by those on top. The ventilation fans were turned on full blast and Holt was brought to the surface and within a few days was said to be fully recovered. But only four months later, on 21 July, James Holt died of 'congestion of the lungs.' It is very likely that the two events were related.

Holt's death did not prevent Marie from continuing his obsessional campaign. Each new County Council was in turn bombarded with letters seeking redress for the imagined wrongs that had been done to him. In the end the council constructed a road up to Holt's pit along the surveyed deviation that was well outside Marie's lease.

William J. Holt, James's son, took over his father's mine and in 1893 added to the feud by accusing Marie of flooding his (Holt's) mine. Marie had diverted a water race into his mine that was on fire, but the Inspector of Mines could find no evidence that Marie's actions had flooded Holt's mine. But the mine was certainly flooded and the shaft had collapsed. Young Holt had to cart coal from Alexandra to satisfy his customers.

The flooding was not the disaster it might have been because Holt's mine was already very much worked out, and before his death, Holt had been planning to open another shaft further down the hill. His son, with George Griffiths as manager, completed this shaft, but the results were disappointing. An extraordinary decision was then made. It was decided to drive a low-level drive into the coal seam from the valley of Dairy Creek. The scheme, which was almost an exact duplicate of that proposed by Marie eight years before, involved tunnelling through about 1,000 feet (300 m) of solid schist rock. The work was financed by a group calling itself the Dunstan Coal Company, which included among others, W. J. Holt and G. Robertson. It was managed by G. Griffiths. The whole operation took over four years, but in the end netted only a few hundred tons of coal before it was abandoned.

That was the end of the 'Waikerikeri' coal mine. Coal production had virtually ceased with the flooding of the mine in 1893. According to official returns, over its 14 year life, it had produced 4,389 tons of coal.

Vincent Mine

Such was the demand for coal from the growing number of gold dredges

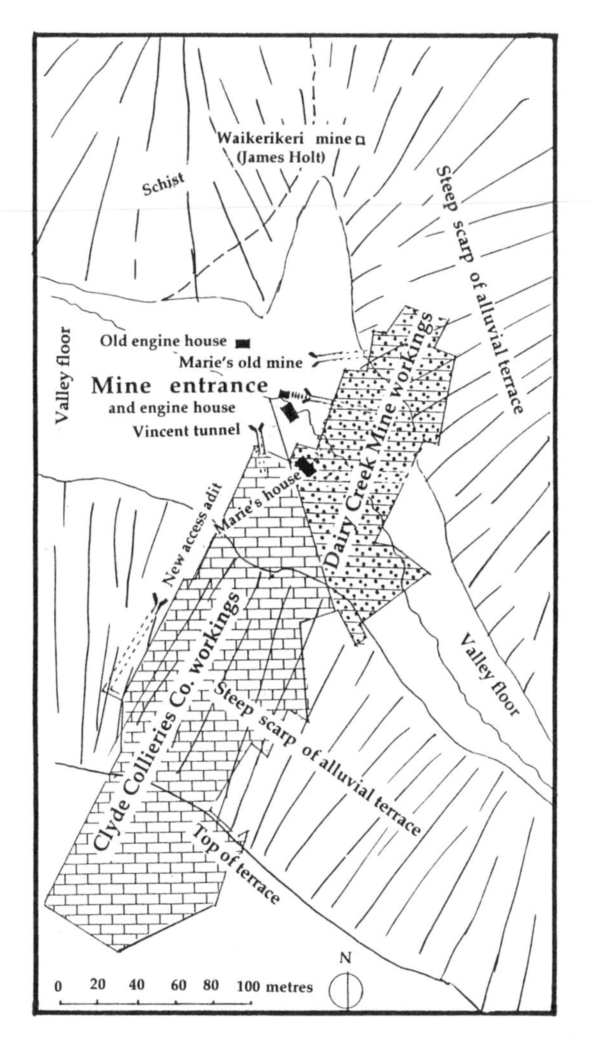

Figure 7. 8. Coopers Gully coal field in 1906 showing ground worked by the Vincent mine (Clyde Collieries Co.) and Dairy Creek mine (W. J. Tonkin). After the Clyde Company took over its neighbour in 1903 the Dairy Creek mine adit became the main entrance to the combined workings. (Adapted from Park, 1906 with additions from mine plans).

that a new company, the Clyde Collieries Company, was formed in 1898. William Kitto, an experienced miner, was appointed to open up a new mine. The 'Vincent' mine was close to Marie's old Dairy Creek mine but ran under the terrace in a southeasterly direction. George Turner was appointed manager in 1899, and assisted by a steam engine for hauling trolleys, almost immediately began producing over 3,000 tons of coal a year. Production peaked in 1902 with 11 men hewing 5,255 tons for the year, by which time the coal was being mined 100 metres from the

Figure 7. 9. Clyde Collieries mine in 1906. Mine house left foreground, Mine entrance, centre, with Marie's old mine immediately behind. The unused engine house in the left background provided haulage for the now unused Vincent adit Holt's old mine is further up the gully beyond the fruit trees which are the remains of Marie's orchard.

entrance.

In March 1903, the Clyde Collieries Company took over the Dairy Creek mine. It was not long before the two mines were joined underground with connecting drives, and worked as one concern under the management of George F. Turner. Less and less coal was taken from the Vincent section of the mine, and by 1905 it contributed only 378 tons. The Vincent adit was no longer used for bringing out coal, although a new tunnel had been driven into the most westerly part of the workings to provide an easier emergency exit. All of the production from both sections of the mine was brought out by way of the Dairy Creek mine which was now the only mine working in the Gully.

In 1905 George Smith took over for a spell as manager, but Turner returned in 1908 to find the demand for coal was falling as the number of dredges steadily declined. The company went into liquidation in 1909, and was bought by one of the mortgagees, Jonathan Rhodes, who continued to work the mine until it finally closed in 1913.

The mines of Coopers Gully — the Waikerikeri, Dairy Creek and Vincent pits —from 1871 to 1913 had between them produced nearly 50,000 tons of coal.[13]

Figure 7. 10. Coopers Gully today, The disputed road up to Holt's mine is clearly visible in the centre of the photograph.

Today

From the top of the nearby terrace, Coopers Gully does not give any indication of the activity of 100 years ago. A row of mature Lombardy poplar trees running up a ridge, and others, together with a few white poplars, scattered about the floor of the valley, are the most striking features. Everywhere brier is rampant. Closer inspection, however, reveals recognisable remains of structures associated with the coalmines.

The road first built by Collins Marie can be easily followed down Dairy Creek until it disappears under the waters of Lake Dunstan. Holt's road, which caused so much squabbling in the early days, can be followed up the gully past a group of old fruit trees which are the remains of Marie's orchard, to a cluster of low stone walls almost hidden by brier bushes. These mark the site of James Holt's coalmine.

On the flat floor of the gully, bricks and stones amongst the undergrowth are all that remains of the boiler house shown in Park's 1906 plan and photograph (Figures 7.8 and 7.9). Among the brier are the remains of a loading bank, and the walls of a stone building, which is marked as the 'Mine house,' and is probably the house Marie built to replace the tent he lived in when he first came to the Gully. A slumped, brier-filled trench, running into the foot of the terrace, is now the only sign of a former drive and almost certainly marks the position of the last access adit driven by the Clyde Company. All other tunnels, including that of the main Dairy Creek mine were filled in long ago.

Figure 7.11. The elongated patch of brier in the centre of the photograph marks a slumped adit which gave access to the mine workings.

OTHER COAL MINES

Outcrops of coal occurred at a number of places around the southern end of the Manuherikia Valley and some were mined.

Blackmans Gully

By the end of the 1860s, all coal mining on the west bank of the Clutha River had been abandoned, and coal for the considerable population

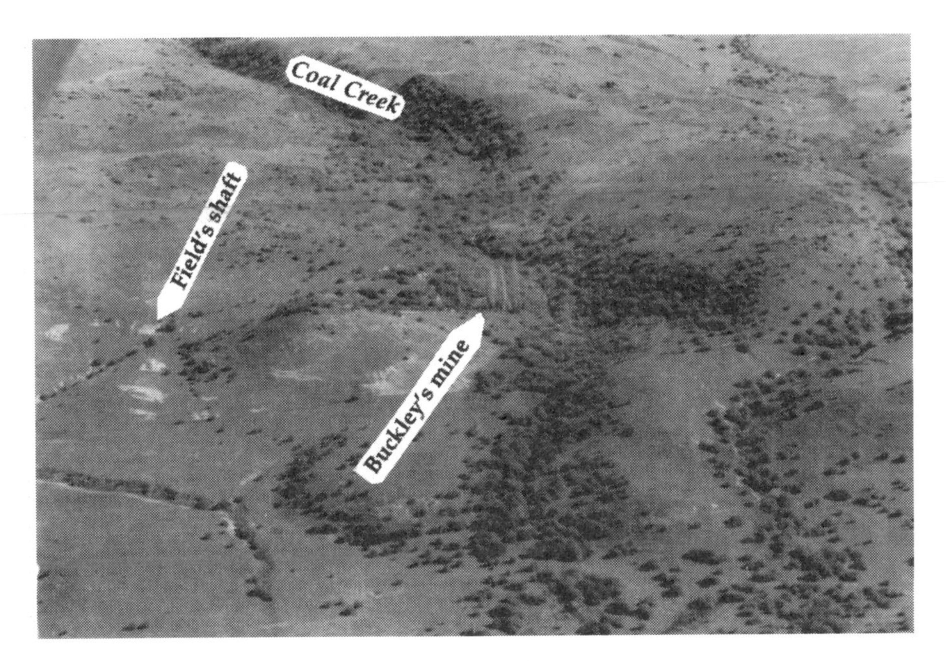

Figure 7. 12. The site of the coal mines in Coal Creek, Blackmans Gully.

residing along the west side the river had to be punted across from Clyde or Alexandra. So when, in 1871, Craven Paget discovered coal at Blackmans Gully, a secluded valley at the foot of the Old Man Range on the western side of Earnscleugh Flat, the local people were delighted. But no mining was attempted for some years.

It was 1876 before Holden and John Buckley began to mine the coal in Blackmans Gully. Their Earnscleugh Colliery was opened up in Coal Creek, a tributary of Omeo Creek that flows through Blackmans Gully. The seam, which sloped down to the southwest, was 14 feet (4.2 m) thick although only 8 feet (2.4 m) was worked. The coal, which was excavated on the bord and pillar system, was drawn out of the mine by horse-drawn trucks through an adit over 30 feet (10 m) long. The partners offered coal at 7s. 6d a ton at the pit head. A fire in January 1882 caused some disruption to the workings and led to the breakup of the partnership, but Buckley carried on.

On 15 March 1888 this mine was the scene of the only fatal accident known to have occurred in a local coal mine.[14] Just the day before the accident, John Buckley had cautioned his stepsons, men of 30 and 27 respectively, about the dangers of going down the 30 foot shaft without first testing the air by lowering a lighted candle. The elder stepson had been guilty of this omission the preceding day. "You will do it once too often," warned Buckley.

The following morning one of the stepsons, James Statham, was lowered down the shaft by his brother, John, without having carried out the candle test. Soon after the man reached the bottom of the shaft he called to be pulled up again. Half way up the shaft he appeared to let go the rope and fell to the bottom. He again grasped the rope and was pulled up a couple of feet before falling again. After repeated calling without effect John ran for assistance. A bunch of scrub was pulled up and down the shaft to clear the foul air before the body could be retrieved.

The Mines Inspector thought that products of slow combustion from a small fire in adjoining old workings had found their way into the mine. No blame was attached to Buckley, the owner, or to the surviving brother. George Field, a local miner and one time proprietor of the Butchers Gully hotel, took over after this but within a short time the mine was closed down.

Nearly 10 years later, in 1896, in the face of increasing demand for coal for gold dredges, an attempt was made by George Field, with partner Hale, to reopen the mine. They sank a shaft, 65 feet (20 m) deep, south of the old workings and reached good coal standing in a vertical seam 25 feet (8 m) thick. They took out a few tons of coal and talked of opening up the mine by driving a tunnel, but nothing more was heard of the scheme.

Shepherds Flat

High up on the flanks of the Old Man Range at an altitude of 620 metres and only a few hundred metres from the present Fraser Dam, an outcrop of coal had been noted by early prospectors. Rumours that a gold dredge might be placed on the nearby Fraser River sent William J. Holt up the mountain in 1901 to sink a prospecting shaft in the shallow gully. It disclosed a seam of coal six feet thick under 16 feet (5 m) of overburden.

When the *Loch Lomond* dredge finally started work in December 1903, near the junction of the Hawks Burn, Holt was ready for it. But he had a great deal of difficulty carting the coal to the dredge especially in the winter when the ground was frozen, and some of Holt's horses actually died from the effects of the cold. Becoming impatient with the delays, the dredging company bought the mine in October 1904 and produced a few hundred tons a year. Access to the coal seam was by way of a sloping drive and a horse drawn trolley brought out the coal. Because there was lower lying ground nearby, the mine could be drained by a siphon.

The dredge closed down in mid-1906 and the coalmine with it.

Gorge Creek

There was a flurry of excitement at Bald Hill Flat when James Austin and his mate, T. Dickson, announced in December 1892, that they had found a seam of coal at nearby Gorge Creek. Austin and party seemed to mean business when they sank a shaft and reached the seam at a depth of 20 feet (6 m) and then applied for a 240 acre Mineral Licence. They got

Figure 7. 13. The excavation at the centre of the photograph marks the Shepherds Flat coal mine. It was opened to supply coal to the *Loch Lomond* dredge which worked in the Fraser River, near the junction of the Hawks Burn, about two kilometres distant.

valuable publicity when they sent the newspaper office some samples which were pronounced of good quality.

But that was the last that was heard of the Gorge Creek coal mine.

Shingle Creek

An advertisement appeared in a couple of issues of the *Dunstan Times* in June 1896, announcing that T. McLoughlin of Shingle Creek was offering prospective customers the best coal in the district for 7 shillings a ton at the pit mouth, 15 shillings a ton delivered at Bald Hill Flat, and further afield by arrangement. There was more. Intending customers could have a free bag for a trial!

The mine, named the 'Black Diamond,' was described by the Inspector as a small pit suitable only for local use. The coal was of poor quality and after trying unsuccessfully to sell the mine, McLoughlin closed it in late 1900.

NOTES

1. *Otago Provincial Gazette*, 7 January 1963.
2. Holt apparently continued to work the mine at Clyde for some time after he had opened the mine in Cooper Gully. It is recorded that he extracted 500 tons of coal over the next few years.

3. So named, according to Gilkison, (1936 p. 46) after the notorious 'Dunstan Dairy,' a sly-grog shanty that sold milk heavily laced with alcohol. Sumpter (1947) records that the remains of the stone walls of the 'dairy' were still visible in the 1940s.

4. Collins Marie was a Frenchman with a fiery temper. It is said that when James Holt and he happened to meet in the main street of Clyde, Marie's shouting could be heard from one end to the other. According to old-timers he was called locally

COAL MINES

NAME of MINE	OWNER (Manager)	PERIOD	OUTPUT (Tons) under each owner
	CLYDE		
DUNSTAN	J. Morgan	1863-1864	
	J. Holt & Dakin	1864-1878 (closed)	15,000
	COOPERS GULLY		
DAIRY CREEK	Collins T. Marie	1871-1887	
		1887-1897 mine idle	
	(J. Pratt)	1897-1899	4,408
	W. J. Tonkin		
	(R Robertson)	1899-1903 (to Clyde Collieries)	6,600
WAIKERIKERI	J. Holt	1882-1891	
	W. J. Holt	1891-1893 mine abandoned	4,389
	Dunstan Coal Co		
	(G. Griffiths)	1894-1898 (closed)	933
VINCENT	Clyde Collieries Ltd		
	(W. Kitto)	1898-1899	
	(G. Turner)	1899-1904	
	(Geo. Smith)	1905-1908 (liquidated)	25,208
	Jonathan Rhodes		
	(G. Turner)	1908-1913 (closed)	7,754
		Total tons for Coopers Gully:	49,792
	BLACKMANS GULLY		
EARNSCLEUGH COLLIERIES	J. Buckley & C. Holden	1876-1882	
	J. Buckley	1882-1890	
	G. Field	1896-1896 (closed)	3.844
	UPPER FRASER		
SHEPHERDS FLAT	W. J. Holt	1903-1904	
	Loch Lomond Company	1904-1906 (closed)	1,433
	SHINGLE CREEK		
Black Diamond	T. McLoughlin	1896-1898	
	P. Galvin	1898-1900 (closed)	232

Figure 7. 14. A summary of the coal mining industry in the district (excluding Alexandra).

'Toosant ' Marie. Even officials had difficulties with his name and in several mining reports he is called 'Mr Collins.'
5. *Otago Daily Times* 30 August 1871.
6. This was a device for lifting water from a river by the force of the current. It was described by the Provincial Surveyor, J. T. Thomson, at a meeting of the Otago Branch of the New Zealand Institute,
7. *Otago Daily Times* 8 and 11 August 1871.
8. This account is taken from Angus, 1977 p. 85.
9. AJHR 1891 I—2 p. 15; 1892 I — 2 p. 7.
10. AJHR 1881 H —14 p. 16.
11. AJHR 1896 C — 3B p. 14.
12. It was John Statham's brother James who was killed by foul air in the Earnscleugh mine at Blackmans Gully in March 1888 (see 'Blackmans Gully').
13. Figures from Mines Statements AJHR
14. AJHR 1888 C—4 p. 16-17.

8.

A MATTER OF PRIORITY

—Sluicing the 'West Bank of the Molyneux'

After emerging from the Cromwell (Dunstan) Gorge, where its former tumultuous passage is now stilled by the huge dam at Clyde, the Clutha River adopts a more leisurely pace as it flows across the southern end of the wide Manuherikia Valley. Between Clyde and Alexandra the river flows as a single, deep stream in a valley bounded by steep gravel banks some 30 metres or so high. Apart from a slight curve opposite Muttontown Gully the river follows a straight course for five kilometres from the outlet of the gorge, before swinging into a series of broad sinuous curves. At Alexandra it leaves the Manuherikia Valley and enters another narrow rocky gorge.

From Muttontown Gully to Alexandra the river was once flanked by a flood plain,*[1] standing two or three metres above normal river flow, which formed a discontinuous terrace along both sides of the river. It was well developed on the inside of the curves but generally absent on the outside of the bends where the current impinged directly against the high river banks. The most extensive area of flood plain was above the junction of the Fraser River, where for several kilometres the valley is over a kilometre wide. Here the broad river formerly broke into several streams that joined and rejoined as they made their way across this basin. At the end of the 19th Century, this basin was to become famous as a most profitable dredging ground.

On the inside of most of the big curves, the steep river bank above the flood plain was replaced by a series of terraces which descended towards the river like a staircase. The lowest terrace was about 10 metres above the flood plain and was generally covered with a layer of silt up to two metres thick.

Early miners were prolific in naming identifiable features along the river. The inside of each curve was named as a 'Point.' Of the dozen such features named between Alexandra and Clyde, not one name survives on

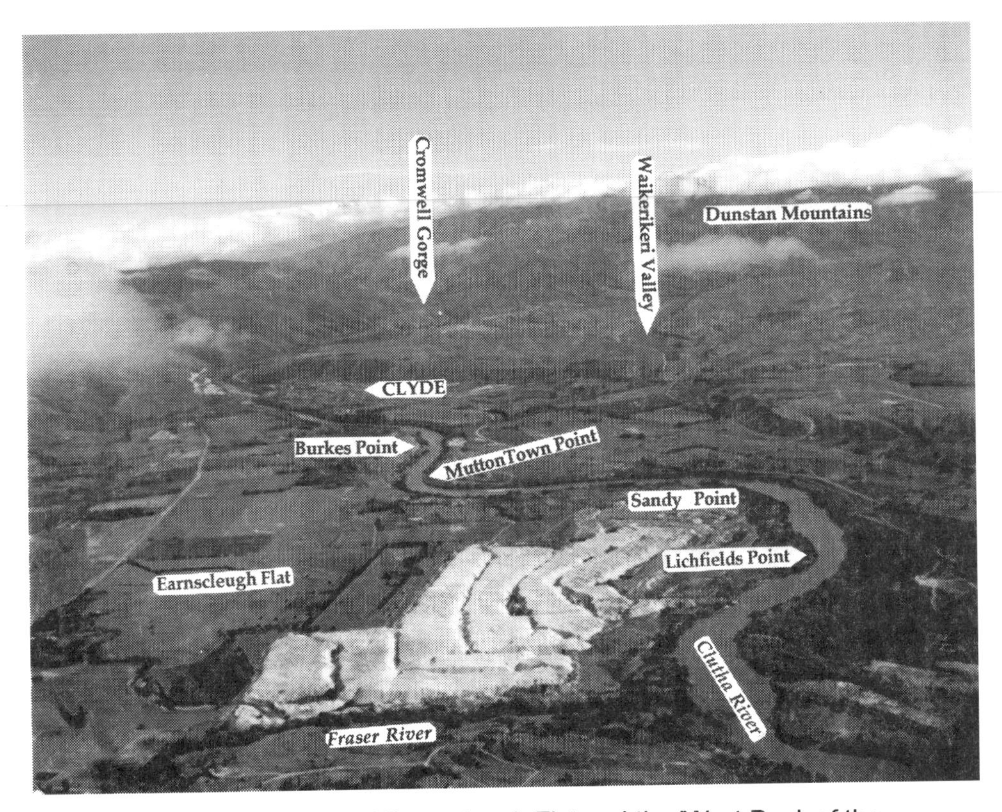

Figure 8.1. Air photograph of Earnscleugh Flat and the 'West Bank of the Molyneux' - the western bank of the Clutha River down stream from the Cromwell Gorge (background). Fraser River joins the Clutha in the foreground. Conspicuous is the large area of dredge tailings created mainly between the early 1950s and 1963. Other prominent features are marked.

modern maps.

Along the western side of the basin above the Fraser River junction the low terrace was particularly wide. The locality was called 'Sandy Point.' At some stage there may have been a clearly defined 'Points'— probably a part of the flood plain round which the river turned to enter the basin. If so, it has been long since dredged away, but it seems that 'Sandy Point' became a general name for several kilometres of river bank stretching downstream from the big bend opposite Muttontown Point nearly to the mouth of the Fraser River.

At the junction of the Fraser River, a fragment of flood plain with a distinctive shape formed by the entrance of the tributary stream, was named 'Sandy Hook.'

Below this junction was a well-formed sandy beach backed by a narrow low terrace. The beach apparently did not yield much gold to the earliest

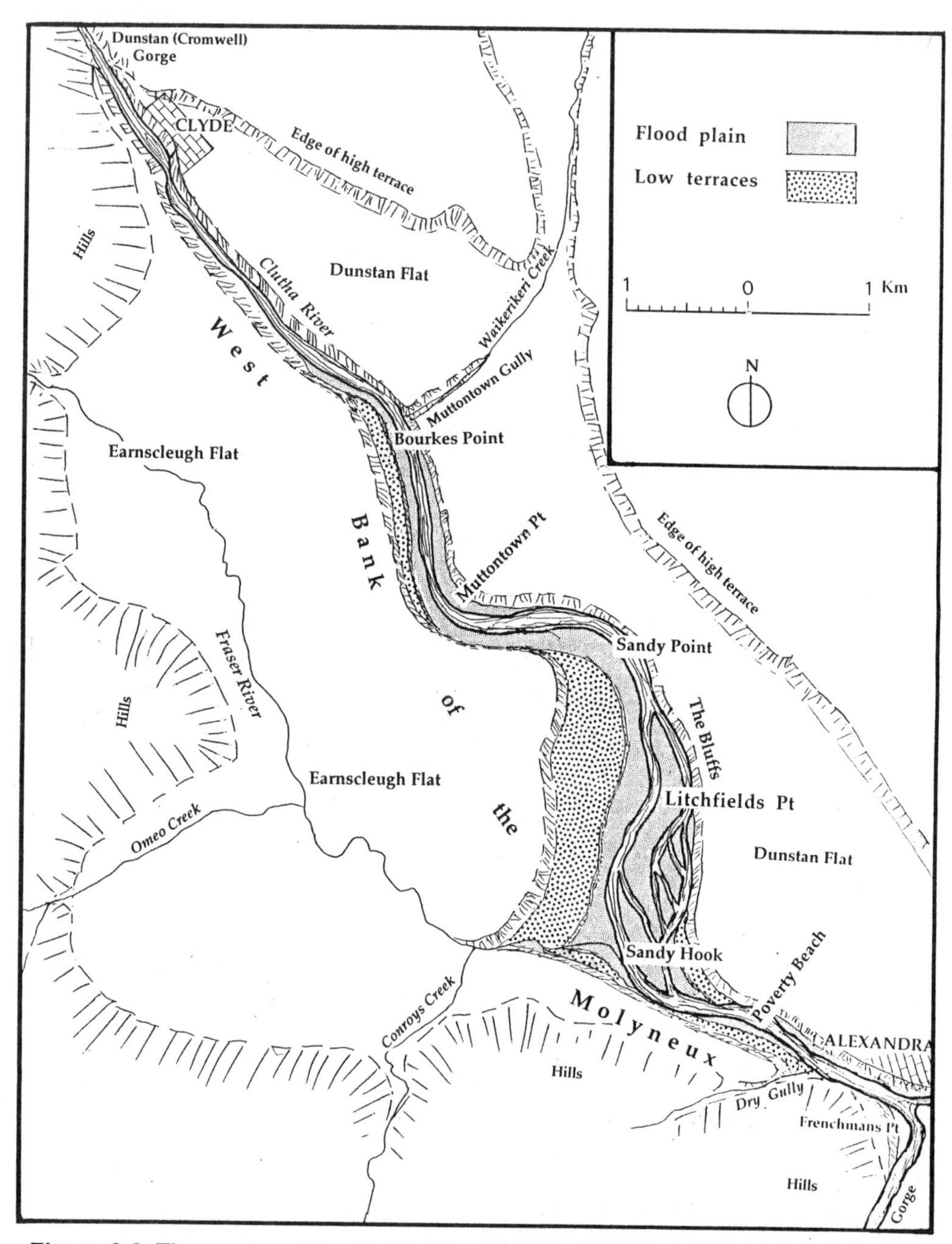

Figure 8.2. The course of the Clutha River between Clyde and Alexandra as it was before dredging. The 'West bank of the Molyneux' is shown, as are miners' names for other localities. For this reconstruction information from old mining maps, air photo-graphs and soil maps was used.

miners so was called 'Poverty Beach,' a name thought singularly inappropriate when, in the 1890s, a large company began sluicing operations. The name was quickly changed to 'Golden Beach'.

'West Bank of the Molyneux'

'West Bank of the Molyneux' was the address of more than 200 miners in 1867. It embraced the western side of Clutha River from Dry Gully (now Chapmans Gully) up to a point opposite Clyde. During and immediately after the Rush the sandy river beaches along this stretch had been thickly populated by miners panning and cradling. The rapidly rising river of September 1862 forced them to seek gold elsewhere. A few of the more experienced miners decided to prospect the low terraces.

Mining techniques required on the terrace were quite different from the simple panning and cradling of the beaches, and were more akin to the methods used at Gabriels Gully. Shafts up to 20 feet (6 m) deep had to be sunk through barren silt and gravel to reach the layer of gold bearing 'wash,'* then the wash had to be carried in sacks down to the river's edge for cradling. The prospects were encouraging, but the labour of stripping away the six to 20 feet of overburden and carting the wash down to the river was daunting.

However, miners with experience of gold mining in California knew how to win gold from these terraces. This was not the place for the lone miner with his tin dish or cradle—what was needed was a group of up to ten miners pooling their resources to bring water to the claim rather than take the 'wash' to the water. So they cut a water race from the Fraser River and used the water to wash away the surface layers of barren silt and sand before washing the gold-bearing sand through sluices.

It was probably this party of Californian miners whom Gabriel Read saw while carrying out an inspection of the newly discovered Dunstan Goldfield for the Provincial Government. He noted[2] about 17 September 1862, only about three weeks after the Dunstan Rush had arrived, that a party had cut a 'little canal more than a mile long' from the Fraser River to their claim near the mouth of the river where they were sluicing the sand off the banks of the Clutha River. This race was probably the first substantial water race on the Dunstan, and this Californian party probably also introduced the method of 'ground sluicing' to the district. Other miners quickly adopted their methods, and were fortunate in that they had a large source of water, the Fraser River, only a few kilometres distant and easily accessible

Fraser River

Rising in a high alpine basin behind the Old Man Range, the Fraser River descends through a series of narrow, steep rocky gorges. It emerges from the lowest gorge and flows across Earnscleugh Flat in a meandering course, more or less parallel to that of the Clutha River before it turns nearly at right angles to join the big river. Its course steepened and included a low waterfall over its final kilometre. It was estimated that in full flow the Fraser River carried about 100 'heads' of water, but like all Central Otago streams its flow fluctuated seasonally.

Water races

Each new party of miners on the terraces had a choice of water supply. They could buy water from the owner of an existing race or they could cut their own race from the Fraser River.

Before a party could construct its own race it had first to apply to the Warden's Court for permission to do so. There was a procedure to follow. The line of the race had to be pegged out and notices posted at both ends of the line. Copies of the notices had to be sent to those already drawing water from the water source to give them a chance to formally object to the application. If the application were successful in the face of objections, the application would be granted and a Certificate of Registration issued. This certificate had to be renewed each year.

At the same time an allocation of water would be made according to the number of miners who were going to work on the claim. For example, a party of one or two miners would receive one sluice-head, four would receive two sluice-heads and then another sluice-head would be allocated for every additional two miners. This allocation of water was a 'water right' and carried a priority, based on the system established on the Californian gold fields of 'first in time, first in right,' or 'first in, first served.' Those who held a high priority water right could, in times of water shortage, demand that all those with lower priority rights stop drawing water. When water races, or shares in water races were sold, the water right and the priority went with the race. Races with high priority rights were valuable assets and were bought and sold for large sums.

By mid-1864 no fewer than seven large races were conveying water from the Fraser River to the West Bank. Over the years, many more races were cut to carry the 14 separate water rights granted from the Fraser River by the end of the century. Old races were abandoned and new ones cut as holders of the water rights moved to new claims or sold the rights. Scores of branch races led from the main races as water was supplied, at a price, to claim owners who did not have their own water rights. Earnscleugh Flat became criss-crossed with a network of water races.

Race cutting from the Fraser River across the surface of Earnscleugh Flat to the West Bank was easy. In fact it was too easy — the loose gravel underlying the Flat was extremely porous, and before the races would carry water they had to be sealed with silt. Then the race had to be maintained by constant patrols to clear out accumulated silt and weeds, to make sure that the race did not leak, and that any breaks in the race wall were quickly repaired.

As with the ownership of the claims themselves, membership of the

Figure 8. 3. The course of some of the old water races are still visible. Here is the trace of the old No 5 race approaching McPherson Rd.

groups who owned races was constantly changing. No 8 race, as an example, was taken over by John Bourke and three others in 1866, and although Bourke continued as a shareholder right through until at least 1900, he had 15 different partners in the ownership of the race over 34 years.[3]

Mining the Terraces

Experienced miners showed how water could be used on these terrace claims to save the hard labour of removing the barren overburden that covered the gold bearing wash. Water was led over the face of an initial excavation where it quickly eroded a narrow channel back from the face. By moving the water, a series of parallel channels were cut and the blocks of overburden between them either collapsed on their own accord or were assisted by men with picks. Work went on day and night, and a flow of 3 or 4 heads of water could wash 1,000 cubic yards (800 m^3) of debris into the river every 24 hours.

When the gold-bearing 'wash,' consisting of a layer of coarse gravel and sand one to four feet (30 cm to 1.2 m) thick, was exposed over an area of a quarter of an acre (about 1000 m^2) a sluice was set up between the exposed 'wash' and the river. The sluice was simply a trench lined with flat stones.[5] Later when timber became available sluice boxes were made from three boards — two sides and a bottom across which were attached

Figure 8. 4. 'Herring bone' tailings on West Bank. The sluice/tailrace is between the left and right sets of tailings.

parallel strips of wood to catch the gold.

A patch of wash (a 'paddock') about 15 feet wide was systematically excavated with pick and shovel, working away from the sluice, not at right angles, but in a slightly upstream direction. This meant that the wash could be shovelled or swept downhill into the sluice where it was washed through the sluice by water from the race. Any gold was trapped under the edges of the flat stones or by the strips of wood laid across the floor of the sluice-box

Stones, large enough to block the sluice, were removed and built into a stone wall about three feet high along the downstream side of the patch of wash. Sand and gravel were washed through the sluice and down a 'tail race' into the river or on to waste ground. As the work proceeded, stones were thrown behind the stone wall as it was extended to keep pace with the work. So when the excavation of the patch of wash was completed it was flanked on the downstream side by a pile of stone held in place by a stone wall running at a slight upstream angle to the sluice.

Operations were then repeated on the other side of the sluice and when this paddock was worked out, the sluice was extended upstream and work began on the next patch upstream. Stones from this were tossed on to the previously worked ground but held in place by the stone wall. This method of mining gave rise to a 'herring-bone' pattern of stacked stones, still clearly evident today,[6] which was characteristic of ground sluicing on the terraces.

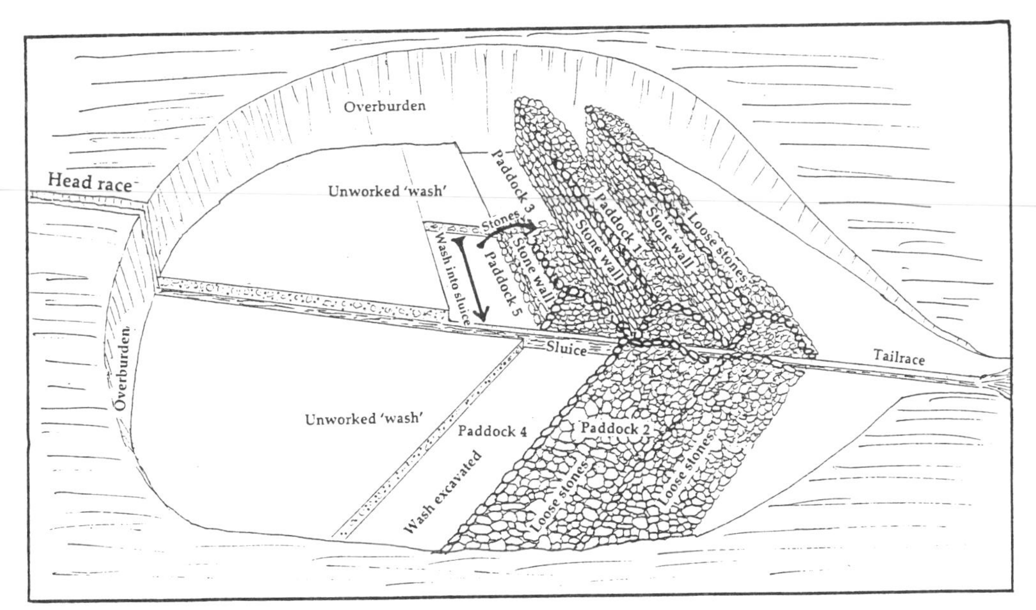

Figure 8. 5. How 'herringbone' tailings were formed. After the overburden was removed, 'paddocks' of wash on either side of the central sluice were excavated alternatively. Stones, too large for the sluice were heaped behind a stone wall on the down-stream side of each patch.

Working with large flows of water could have its exciting moments. One night the wall of a race carrying six heads broke away, and before the miners were aware of what was happening, a chasm 200 yards (180 m) long, 50 feet (16 m) wide and 40 feet (12 m) deep had been excavated in the alluvial terrace.

The *Otago Daily Times* [4] was impressed with this method of large scale sluicing, pointing out that systematic working of large areas was much more successful than trying to pick out bits of rich ground for working. Many men who had 'been running here and there all over the Province' were at last returning to engage in this more secure form of mining. One of the problems, though, was the small area of the claim allocated to each miner. The size of claims had been established based on beach claims suitable for one man working with a pan or cradle. Sluicing a small claim was not worth while—the cost of bringing in water alone would not be covered by the return of gold.

The paper went on to predict that once the regulations had been changed to allow a sensible size for sluicing claims — claims that would take at least 12 months to work out instead of six weeks, then men would settle down and build comfortable stone houses instead of living in tents:

One thing must be borne in mind—that a population continually carrying their [*sic*] blankets upon their backs, can never make a country rich and prosperous.'

Figure 8. 6. Retaining walls of stones on the upstream side of each pile of tailings.

Sluicing the West Bank

By mid-1863 a large race was carrying the first officially granted water right (No 1 priority)[5] from the Fraser River to the West Bank. It was said to be supplying a 'head'[6] of water, presumably in some sort of rotation, to each of 20 claims at Sandy Point. These claims made up what the newspaper described as the 'most systematic works on the Dunstan' and for a time the area was referred to as 'Sluicers Point.' As word of high returns spread, and reports of £10 a week per man were not unusual, other groups were formed, and after being granted their water right and cutting their race from the Fraser River, they settled into the work. Not all were successful. One group led by Mark Cullen took out a claim at Sandy Point in July 1863, which they later called the 'All England Eleven'[7] after the first visiting English cricket team. They cut their race (No 4) from the Fraser River, and when they finally struck gold were reported in 1865 to be making £30 week for each man. But they had bought a great deal of expensive gear, including a large water wheel and Californian pumps, and found themselves in debt. In early 1866 the claim and race was taken over by a group of Clyde businessmen who formed a limited liability company.[8]

Further down river was the Albion group, led by William Fitzgerald, who in mid-1863 took up a claim near the mouth of the Fraser River. This was the only one of the early West Bank claims to be surveyed and have a

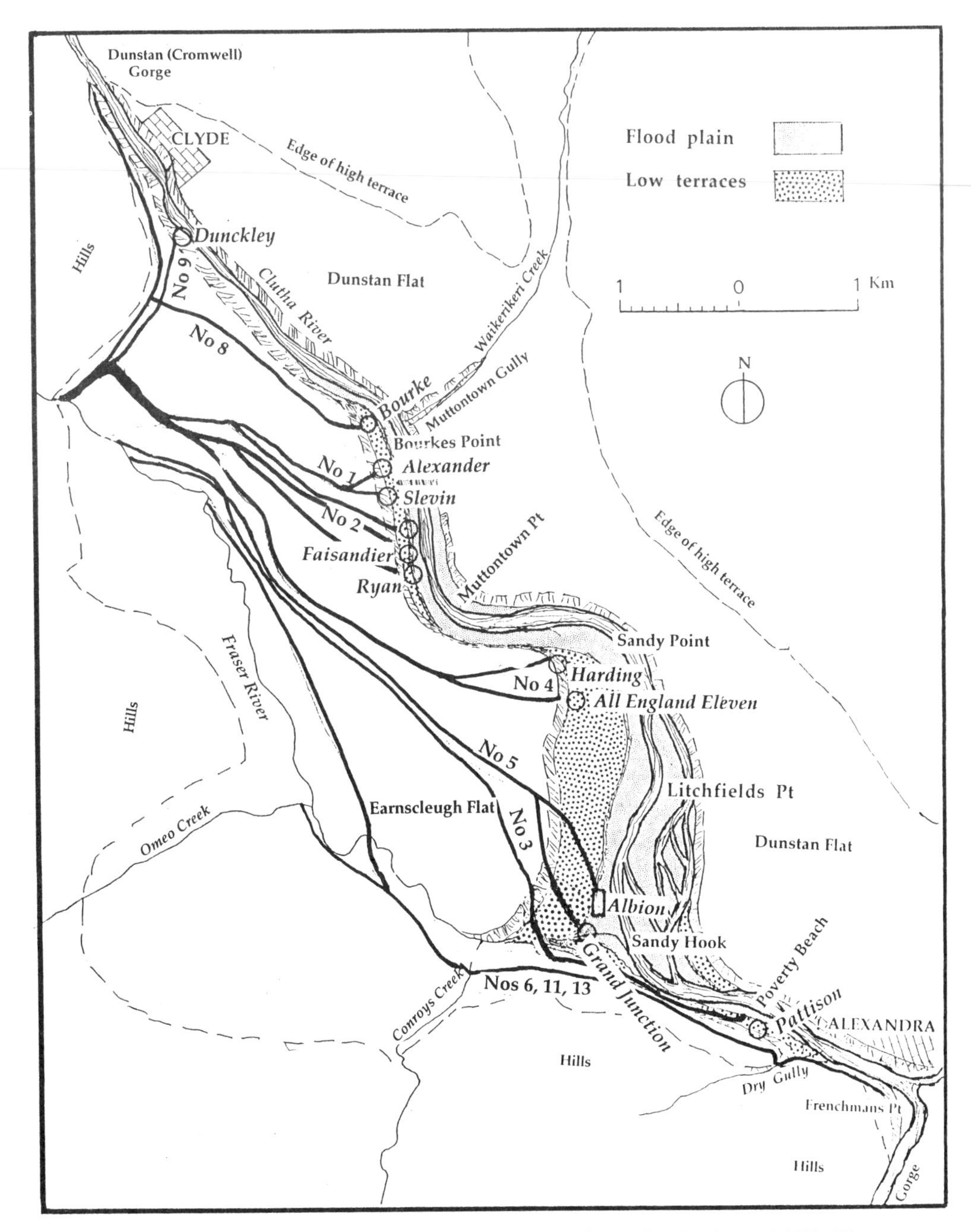

Figure 8. 7. Water races serving claims on the West Bank about 1870. The numbers refer to priority of water right. Claims are named after the principal shareholder. The only claim with a surveyed plan was the 'Albion.' The locations of the others are based on descriptions.

plan prepared. A large overshot water wheel 24 ft (7.3 m) in diameter, said to be the largest in the province, was installed a couple of years later. It was built in Clyde, carried down to the river by 30 men and floated down to the claim. It was turned by water from race No 5 and used to drive Californian pumps to keep the claim dry.

A miner who later became a well-known farmer and fruitgrower on Earnscleugh Flat, Claude Felix Faisandier, took up a claim just above the Fraser River junction. Soon afterwards he acquired No 2 water right and in the early 1870s moved up the river to a new claim. Although still within the general area referred to as 'Sandy Point,' it was about two kilometres up stream from Sandy Point itself. He was one of those who gradually established an orchard on the West Bank, but still retained his interest in mining until he died in 1916.

Several parties had established themselves on the low terraces at this new centre of mining activity that was just down stream from the slight promontory, later to be named 'Bourkes Point.' Furthest upstream were Alexander and party who had invented an interesting pump that was a forerunner of the hydraulic elevator. A pipe 20 feet long and six inches in diameter was erected on supports in a slanting position over the pond to be drained. Another smaller diameter pipe led from the lower end of the long pipe to the bottom of the pond. Canvas hose connected to the bottom of the sloping pipe discharged water at high pressure up the pipe. By a 'well-understood principle in hydraulics' a vacuum was formed and water was rapidly sucked up the smaller pipe from the pond.[9]

Ryan and party, just below Alexander's party, settled in in 1866 for what turned out to be a long period. In 1896 Ryan was heard to remark that he had been at Sandy Point for 30 years and thought it might be time for a change.

James McDouall and party, with water right No 8, had cut a race in 1864 from the mouth of the Fraser gorge to their claim nearly a kilometre above Clyde. It was this water right that John Bourke, William Jamieson and William Richards took over in 1866 and brought the water in a race along the bank of the river to what became known as Bourkes Point, opposite Muttontown Gully. Here they were joined by James Slevin[10] who had opened a claim on the narrow low terrace. Slevin had had a small share in No 1 race since 1863, but in 1872 he became principal shareholder, and with Daniel McElroy, William Ryan and Williamson as partners, diverted the race to Bourkes Point.

Edmund Jones, Water Race Entrepreneur

Edmund Wellington Jones had arrived at Dunstan with the Rush. He was presumably a man of some means for he began to invest in water races. He quickly formed a partnership with Knowles and Thomas Oliver who, at the time, were owners of the No 1 race to Sandy Point, and then he went on to acquire interests in other races supplying the West Bank. But the West Bank was not his only interest.

Jones constructed the 'Mountain Race' from the south branch of Conroys Creek that was able to deliver water, by way of Chapmans Gully, to Frenchmans Point just as fabulously rich gold was discovered. Later he cut races that brought water from the Fraser River, Omeo Creek and Conroys Creek into this race, so he was able to supply seven heads of water to Frenchmans Point at a weekly charge of £10 for each head. He built a large six-roomed stone house[10], which he called 'Como Villa' and purchased the adjoining Grange Farm. Unfortunately he died of a heart attack in April 1868 and his interests in his various races passed back to Oliver and Knowles. Among these was the valuable first water right (No 1) which was now supplying water to no fewer than seven companies on the West Bank. They also had No 3 race that supplied claims at Sandy Hook.

By early 1864 several groups had taken up claims at the mouth of the Fraser River, especially on the low terraces just upstream from the mouth where a protruding spit of gravel gave the name the 'Sandy Hook' to the locality. A branch race was cut from the big race supplying Sandy Point claims but it was not long before a group decided that Sandy Hook miners needed a race of their own. They began to cut a race (No 3) from the mouth of the gorge of the Fraser River and used a plough (an innovation at the time) to mark out the line of the race and to speed things up. This race was bought by Edmund Jones and left to Knowles and Oliver.

Sandy Hook

Using water from this No 3 race, Joseph Knowles and Thomas Oliver took up a claim at Sandy Hook in early 1869, and when James Simmonds joined the partnership in 1870, they called themselves the Earnscleugh Grand Junction Mining Company. With seven employees, the party began to excavate a large 'paddock' in an endeavour to reach gold lying on the 'bottom,' which they hoped was not too far below because water was seeping into their hole in increasing quantities. The Hit or Miss party that had lately finished mining at Frenchmans Point, joined the Grand Junction party and brought with them a tramway and other gear. Now with eleven men, they carried their excavation down 14 feet (4 m) below the level of the river, and using a Californian pump turned by teams of two men working in short shifts, they made a desperate attempt to reach bottom. In fact just before they were overwhelmed with rising water they were able to drive a crowbar down and touch rock bottom three feet below. But that was as far as they got.

Two years later, with a battery of Californian pumps driven by a large overshot water wheel, they made another attempt to beat the water. It was in vain. Simmonds had already gone on to work at Butchers Gully and Bald Hill Flat, but Knowles was ruined in health and finance as, in some

Figure 8. 8. This 1949 air photograph of the junction of the Fraser and' Clutha Rivers, shows herringbone tailings before they were destroyed by dredging.

way, he had lost his share in the races he had inherited from Edmund Jones. He spent his last years as a raceman on the Caledonian race and died in June 1875.

Poverty Beach

At the lower end of the West Bank, between the mouth of the Fraser River and the mouth of Chapmans Creek, lay Pcverty Beach. It apparently belied its name, as Louis Gards and party, well equipped with a large water wheel, along with parties led by John Pattison (No 3 and No 11 races) and by John Young, were recovering satisfactory returns from their respective claims.

Depression

It would seem that 1873 was the busiest year for the West Bank, with over two dozen sluicing claims at work. The most easily worked ground, however, was fast running out and miners were drifting away to other gold fields. It was the beginning of a long depression in mining affairs around the Dunstan. A sign of the times was the advertisement in late 1873 to let out the All England Eleven race and claim. This was followed during the next year by a similar offer to let the Albion claim and race. Partners wanting to pull out of various groups offered their shares for a song, but there were few takers. In 1875 Felix Faisandier offered for sale his second right of three heads from the Fraser River but did not get a buyer. By April 1876 there were only four or five sluicing companies left and these were mostly owned by men such as McElroy, Bourke and Faisandier who were establishing farms nearby.

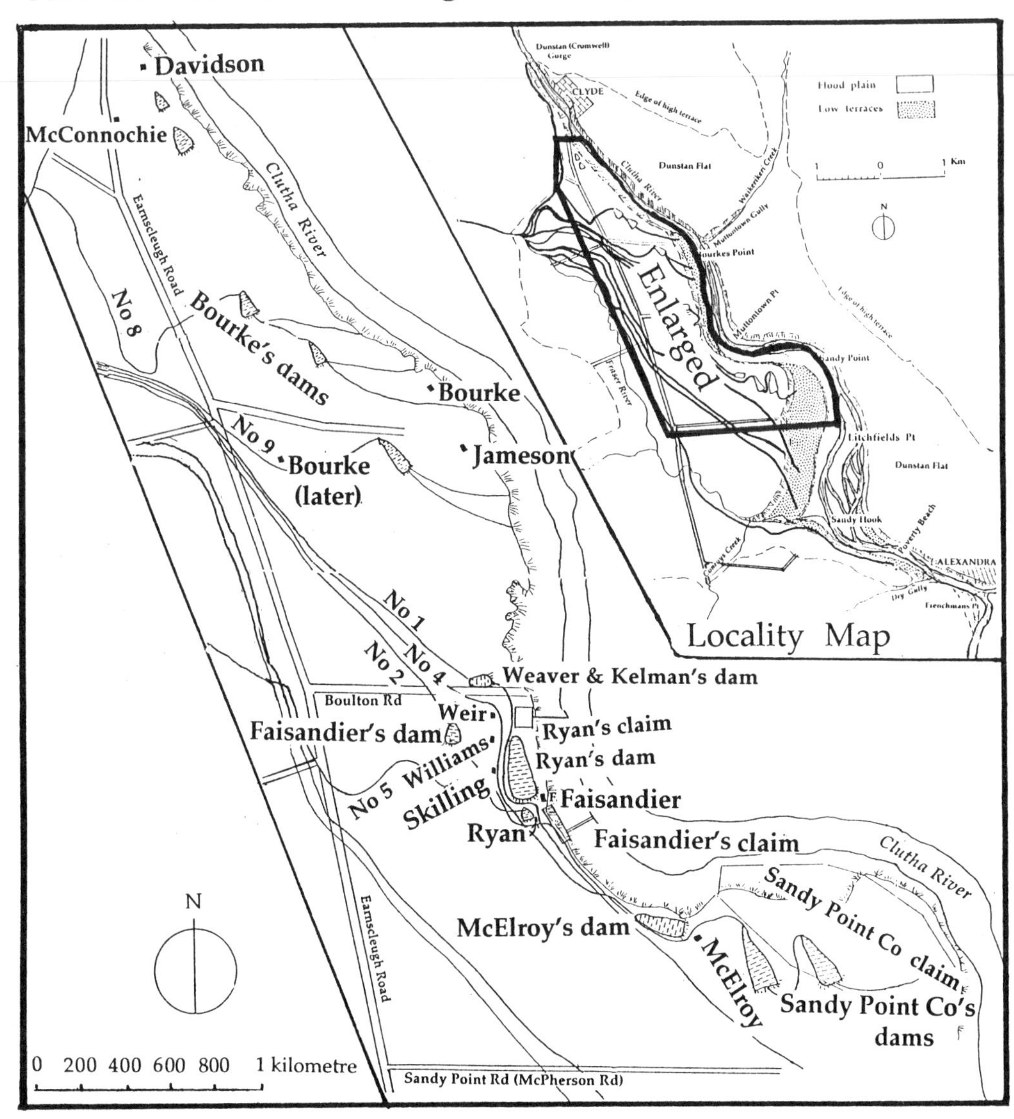

Figure 8. 9. Water races and dams serving claims on part of the West Bank about the turn of the century. Location of races, dams, claims and miners' residences based on descriptions and old mining plans.

Revival

It was not until 1890 that there was a revival when the Sandy Point Gold Mining Company was formed. The shareholders were James Simmonds,

Thomas Hawley and Thomas W. C. Hawley. The claim was a large one of 21 acres (8.5 h) and the plan was to work the claim by hydraulic sluicing with water from No 1 race. Patrick Weaver had replaced James Slevin, who had died in 1887, in the partnership that owned this race.

William Jamieson, who for many years had been a partner with John Bourke and William Richards in the No 8 water right, took over the No 5 water right, rented the old Albion race and began work at Bourkes Point.

There was also much activity downstream at Poverty Beach. Old timers such as William Noble, John Young, Paget and Party and John Pattison were all still working, but they were joined by a new company, the Golden Beach Hydraulic Company that opened for business in May 1896.

Dredges

However, the days of sluicing companies on the West Bank were numbered. Already steam gold dredges such as *Perseverance*, *Enterprise* and others were chewing into the flood plain of the river. Within a few years it had been completely dredged. The small early dredges were followed by much larger and powerful machines fitted with elevators which could stack tailings in great heaps behind the dredges. Fitted with these elevators, the three dredges of the Earnscleugh Company, the *Glasgow* dredge of the Sandy Point Gold Dredging Company and the *Matau* dredge, were able to dredge into the lower terraces, and within a few years the workings of the old sluicing companies were dredged through and largely destroyed. Luckily, here and there, small patches of the characteristic herring-bone pattern of tailings remain to give some indication of the extent of these old workings.

The dredging era too, saw the purchase of most of the water rights by the Sandy Point Dredging Company and the Earnscleugh Dredging Company. The many races were consolidated into two or three major races that conveyed large amounts of water to replenish the pond in which the dredges floated. As the vessels worked their way further into the terraces, the pond water tended to drain away and had to be constantly topped up. The situation in 1915 was:

Sandy Point Gold Dredging Company held water rights Nos 2, 4, 5, 10

Earnscleugh Gold Dredging Company held water rights Nos 1, 3, 6, 7

Bourke and Foulds held water right No 8

Holt, Naylor and Davidson held water right No 9

Four water rights of lower priority were held as follows:

2 by Felix Faisandier; 1 by the Crown; 1 by Earnscleugh Gold Dredging Company.

Irrigation

It was first proposed during a hearing of the Land Commission in May 1920 that the Government should purchase the Sandy Point Gold

Figure 8.10. Three mining episodes on the West Bank of the Molyneux; illustrated by three sets of tailings. Centre foreground: remnant of the once extensive herringbone tailings from the early ground-sluicing of the low terrace. Right: tailings from the early dredging of the flood plain. Left: tailings from dredging of Earnscleugh Flat by the most recent dredge.

Dredging Company's water rights from the Fraser River for irrigation. The suggestion was taken up, and by the end of 1921 the scheme was under construction. Then in March 1924 the Minister announced that the water rights of the Earnscleugh Gold Dredging Company, amounting to some 30-40 heads had been purchased. With purchases of other rights the Government now owned all of the water rights from the Fraser River, and was eventually able to supply irrigation water to some 3,000 acres (1,200 h).

Although the extensive sluicing operations destroyed substantial areas of the low terrace along the West Bank, the effect was small compared with the wholesale destruction caused by the big dredges of the latter dredging era. Even this was eclipsed by the work of the huge dredge of the Clutha River Gold Dredging Company. From the mid-1950s until it closed down in 1963, this machine not only virtually demolished the terraces at Sandy Point but cut deeply into Earnscleugh Flat itself.

Since the demise of the dredges, the water of the Fraser River has been devoted to the irrigation of hundreds of hectares of stone fruit orchards and pastures. The produce from these has exceeded by hundreds of times the wealth produced by the gold mining.

NOTES

1. The flood plain along this stretch of the river was completely destroyed by dredging during the early part of the dredging era.

2. *Otago Daily Times* 10 October 1862.

3. National Archives AAJQ 4/8 125b.

4. *Otago Daily Times* 21 November 1863.

5. The race which carried the first priority water right was known as 'No 1 race.' Similarly No 8 race carried the water granted as eighth priority right. A difficulty in identifying races arises because water rights were sold separately from the races. For example, the holder of No 2 water right running his water in No 2 race might sell his water right to another miner several kilometres distant. The new owner would divert the water into a new race or into an already existing race which now became 'No 2.' race. The original No 2 race would be abandoned or perhaps used by another claim-holder and used for conveying water with a different priority number.

6. 'A Government sluice-head' (abbreviated to 'head') was equivalent to a flow of one cubic foot per second running for 12 hours. Later regulations increased this to 24 hours. Each mining party quickly built a dam on the terrace high above its claim, and its allocation of water was delivered to its dam.

7. The All England Eleven was the first international team to visit New Zealand. The team, under their captain Parr, arrived at Port Chalmers in January 1864 in the midst of a storm. Their coaches were much delayed by fallen trees on the road, which at the time led by way of Upper Junction to Northeast Valley. They were greeted by crowds lining the road. A four-day cricket festival was held with players from all over the province invited to take part and to compete for a place in the team to play the Englishmen. Information courtesy of George Griffiths, cricket historian.

8. The claim and race was taken over by a group of investors with John Cope, as Chairman, and James Hazlett, Goodwin, Joseph Hastie, William Theyers, W Grindly and Cox E Aldrige, as shareholders. J. V. Cambridge was manager.

9. *Dunstan Times* 26 May 1871.

10. There is variation in the spellings of this surname. It is recorded as 'Slevin' in the Electoral Roll of 1880-81 but 'Sleven' in 1884-85. In the Burial Register of Clyde Cemetery it is 'Slaven,' but on the headstone it is spelt 'Slevin,' the version adopted here.

11. This house was occupied as a dwelling into the late 1950s. The exterior walls, somewhat modified, still exist and are roofed over. The building is now used as a barn. This is almost certainly one of the oldest buildings in the district.

9.

JEAN DESIRE FERAUD

—An Introduction

Jean Desirè Feraud came from the south of France.[1] His home village of Gattieres, where his father Jean Louis, was a doctor and Mayor, was a short distance from Vence, a town only a few kilometres from the city of Nice. Feraud was born in Vence on 13 December 1820, but it was not until 1855 that he comes to our notice as owning a restaurant and bakery in Ballarat, Victoria. He arrived at the Dunstan in 1863 and was sufficiently established in the community to be elected to the committee of the subscribers to the Dunstan District Hospital on 28 January 1864. However, Feraud's membership lasted only a little over two months before he resigned over difficulties 'that existed through misunderstandings in the internal management.'[2]

Figure 9. 1. Jean Desire Feraud.

Mining Investor

In June 1864, when rich gold was discovered on the point opposite Alexandra, the *Dunstan Times* reported:

One of our leading tradesmen here Mr Ferrand (sic) had netted over one hundred pounds as his share for the last three weeks and that clear of all expenses.[3]

Feraud is often credited with making the discovery himself, but there is no record of his ever having done any serious 'hands-on' mining in New Zealand. It is more likely that the strike was made by his fellow-countrymen, Jacques and Theo Bladier, and that Feraud was their financial backer. The brothers, who had been previously mining on Manorburn Flat, came from the town of Gourdon, only 10 kilometres or so from Vence, and were almost certainly well known to Feraud.

The 'Frenchman's Claim,' as it was called, turned out to be one of the most productive in Otago and gave its name to the locality—Frenchmans Point.[4] Feraud must have been a difficult man to work with as after only a few months he fell out with his partners, and according to the custom of the time, the claim was auctioned. It was bought for £455 by Feraud himself, who promptly formed a new partnership and employed a very capable miner, Michael Kett, as manager. Over the seven years of its existence the claim was estimated to have produced about 12,000 oz of gold, worth in today's currency over $NZ 7,000,000. Feraud, the principal shareholder of four partners, must have become a very wealthy man.

When, in 1870, Feraud had a disagreement with his manager, Kett went off to work his own claims, first at Halfmile Beach and then in Golden Gully. The Frenchman's Claim was then taken over by the neighbouring Hit and Miss company. Feraud, with working partner John McQuillan, opened up a new claim at Halfmile Beach, but through shortage of water the enterprise failed, as did an investment in a supposed gold-bearing reef in the hills behind Clyde.

Investing in gold mines was only one of Feraud's activities. By 1864 he had set up a cordial manufacturing business in Clyde's main street calling it the 'Shamrock Store and Brewery,' although there is no evidence that it was a brewery in the modern sense. He also began developing a market garden and orchard on a block of land on Dunstan Flat about three kilometres east of the village centre and across the road from the 34 acre (14 h) block already acquired by the Bladier brothers.

Bladiers' Garden

The Bladiers, with their experience of market gardening and vine growing under irrigation in Australia, became interested in setting up a similar enterprise in Central Otago. They decided, however, that they would continue mining until the proposed garden began to give returns. They selected a block high up on an old alluvial fan that had spread out over

Dunstan Flat from the mouth of the Waikerikeri Valley.[5] The soils were mainly deep sandy loams with a belt of shallower soils running across the block. At this time. early in 1863, the land would be part of Moutere Run, but whether or not the Bladiers sought permission from Shennan, the runholder, is not known. In May 1863, only nine months after the Dunstan Rush, Jacques Bladier applied to the Warden's Court for a water race which he described as:

> . . . commencing in the Waikerikeri Creek a quarter of a mile above the garden six miles from the Upper Township [Clyde] continuing south along the valley and terminating one mile below the garden. Purpose: irrigation[6].

Apart from the fact that the distance from Clyde is greatly exaggerated, several other interesting points are raised by this description. First, the position of the head of the race. As James Holt had already effectively captured all of the available water from Waikerikeri Creek with his prior right of two heads, Bladier placed his intake to catch water coming in from Waipuna Creek, a tributary of the Waikerikeri Creek fed by numerous springs. Secondly, it shows that Bladier had a garden in existence in May 1863. Thirdly, using the garden as a reference point for positioning the race almost certainly means that the garden was on the side of the stream, and the fact that it was described elsewhere as 'adjoining' Feraud's block almost certainly places it across the present Youngs Lane on Sections 55 to 57.

The purpose of this water race was stated as 'irrigation,' but its quoted length of one mile, means it terminated at the banks of the Clutha River where the water could be used for mining. Perhaps the Bladiers had in mind moving their mining operations from Manorburn Flat to the bank of the Clutha River and so be closer to their garden. Or it may have been simply a device to disguise the use of the water for irrigation, which at that time, would be looked on with disfavour. Jacques gave 'Kilgour Hotel, Clyde' as his address for service, and it is possible that the brothers stayed at the hotel while working on their garden.

The Bladiers had irrigation water on their property for the 1863-64 summer, as the matter was raised at a public meeting (see next chapter) in August 1864 when James Holt made sarcastic reference to the 'extravagant use' of water by Bladier on the porous soils of his property. Used extravagantly or not, the supply of water from Waipuna Creek was completely inadequate for the many fruit trees the Bladiers and Feraud, who had leased the adjoining block, had planted out during the winter of 1864. They simply had to have more water for the coming summer.

The Agricultural Reserve

Under pressure from the many miners who wanted to settle on the land,

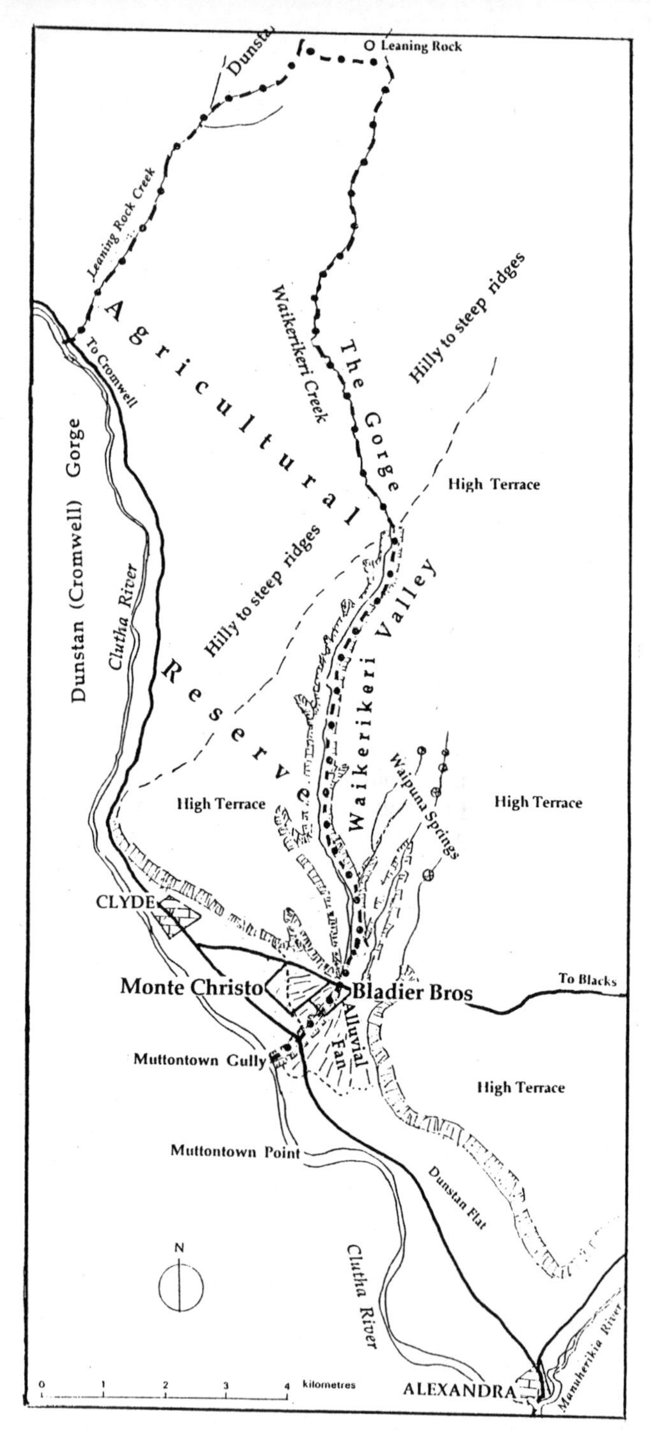

Figure 9. 2. Map of the Agricultural Reserve with Feraud's 'Monte Christo,' the Bladiers' neighbouring property, Waikerikeri Valley and Waipuna Springs.

the Provincial Government decided to declare a large block of land an 'Agricultural Reserve.'[7] A triangular-shaped tract of some 12,000 acres (4,850h) between Leaning Rock Creek in the Cromwell Gorge, Waikerikeri Creek, and the Clutha River, was separated from Moutere Station at a cost of £1,200 compensation. Land suitable for intensive farming was surveyed into small sections in March 1864.

Included in this Reserve was the top of the high terrace (later to be designated as the 'Clyde Commonage') lying immediately north and east of Clyde, but the greater part of the Reserve lay on steep spurs leading up to the summit of the Dunstan Mountains. Locals thought the whole idea was a disaster, pointing out that, of the huge area set aside, only 160 acres were suitable for cultivation. These comprised the flat floor of the Waikerikeri Valley and part of the alluvial fan at the mouth of the valley.

Waikerikeri Creek rises below Leaning Rock, a prominent landmark on the summit ridge of the Dunstan Mountains, and flows south down through a rugged valley that narrows to a gorge, with the stream flowing between steep rocky slopes only a few metres apart. From the gorge, the creek flows for seven kilometres towards the Clutha River in a flat-floored valley, half a kilometre wide and about 20 metres deep, cut into the high terrace of the Manuherikia Valley. In time past this valley had opened on to the then flood plain of the Clutha River, and it was during this period that the creek built an alluvial fan out on to the flood plain.

However, for some reason, probably a change in climate, the Clutha River began to lower its bed leaving the former flood plain as a terrace complete with the old alluvial fan. The Waikerikeri Creek was then forced to cut down through the old flood plain to keep pace with the dropping level of the river. It managed to do this only by markedly steepening the last one and a half kilometres of its course.

So the course of the Waikerikeri Creek is in three distinct parts: the upper valley and gorge, the flat-floored valley through the high terrace, and the steeper part leading through Muttontown Gully (so-called because sheep were slaughtered nearby to feed the miners of the Dunstan Rush) to the Clutha River.

Feraud takes up Land.

A clause in the 1862 Mining Act allowed sections in this agricultural reserve to be leased. Miners did not regard these leases as an immediate threat to their activities as it would still be possible for them to mine the land if the presence of gold could be demonstrated.

Jean Feraud, perhaps with the encouragement of his friends the Bladiers, leased a block of eight sections totalling 100 acres (43 h),[8] on the old alluvial fan just across the lane from Bladiers' garden. Feraud's block contained a higher proportion of stony ground than Bladiers' and this probably accounted for its much larger size. Feraud continued to live in the town, tending to his cordial business, while he planted out and developed his property.

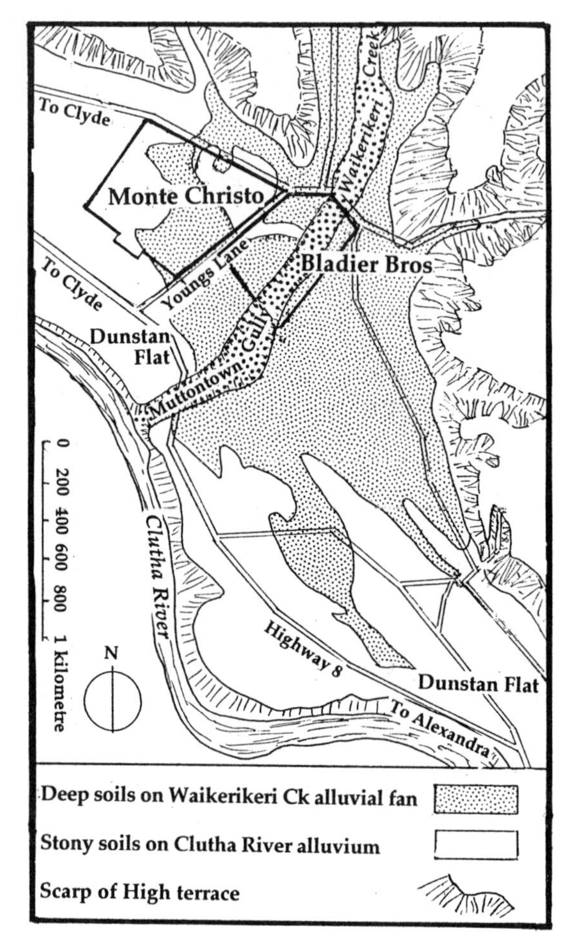

Figure 9. 3. A map of the alluvial fan deposited by the Waikerikeri Creek over what was the stony flood plain of the Clutha River. Deep soils developed on the fine sediments of the fan. Bladier brothers' block had more deep silts, in proportion to its area, than did' Monte Christo.'

By the spring of 1864 Feraud and Bladier had 40 acres in crops between them, including hundreds of vine cuttings which they had obtained from Australia. It was urgent that they should have adequate irrigation water for the coming summer.

There are no further references to the Bladiers after March 1865 so it is assumed they left the district shortly afterwards.[9] Perhaps it had something to do with the dispute between the partners in the Frenchmans claim which occurred at about this time. There is no evidence that their garden was taken over by Feraud so presumably it was abandoned. Their land was later re-leased to Adolphus Oliver who freeholded it in 1871.

'Monte Christo'

Feraud's 43 hectare (100a) farm, 'Monte Christo Gardens'[10] as he called it, flourished. From 1865 he had begun to lease carefully chosen blocks of sections in the Waikerikeri Valley and over the next five years acquired about 200 acres (80h) for dryland farming. As early as January 1866 'Monte Christo' was advertised as in 'beautiful order and open to visitors,' yet the following year the whole of Feraud's properties were up for sale by auction. The Notice of Sale was quite detailed.[11] One hundred acres held under lease from the Crown, comprising 60 acres fenced, with 50 acres of oats and 10 acres of vineyard orchard and market garden. There were also 40,000 plants for sale including young fruit trees, forest trees, shrubs and perennials. The sale included a block of leasehold in the Waikerikeri Valley, also of 100 acres, with 70 acres fenced and under crop. The property did not sell and Feraud continued to develop it.

Montechristo Clyde November 5th 1877.

to the Warden, of the Dunstan District
at Clyde;

Sir.

Take Notice that I do object of granting the application dated october 13th 1877; which I have seen advertised in the Dunstan Times, made by John Mac Nally, Joseph Haste John Lindsey, John Lawson, walter anderson, alexander Mac Leod, and ten others miners residing at the wai-Keri Keri, to the use of the natural bed of the wai Keri Keri Creek for a main tail race, intended to commence at a point

J. D. Feraud

Figure 9. 4. It has been said (e. g. Bodkin, 1998) that Feraud always spelt the name of his farm as 'Monte Christo.' However, in at least one document (part reproduced) 'Objections to mining applications,' clearly written and signed by Feraud, seems to be headed 'Montechristo.'

In 1868, Feraud took advantage of another clause in the Mining Act which allowed a lessee who had held a lease for three years, and had carried out a certain amount of development, to apply for the freehold of the land. Under this clause he freeholded 38 acres (15 h) of 'Monte Christo,' while still holding the remainder on lease. At the same time he applied to freehold 20 hectares of his leasehold in the Waikerikeri Valley. In later years he would freehold another 36 acres (14 h) of 'Monte Christo.'

A reporter who inspected the farm in February 1869 said it contained 800 fruit trees, 4,000 gooseberry and currant bushes and upwards of 2,000 forest trees. In addition there were 200 vines just beginning to bear, asparagus beds which had yielded 100,000 sticks and onion beds yielding 12 tons to the acre.[12]

By 1871 the grapes were yielding so well that Feraud erected a substantial stone winery[13] capable of holding 200 hogsheads. As well as wines he intended to make liquors and cordials. In April 1872 he was able to supply the office of the *Tuapeka Times* with samples of ducal grape wine described as 'possessing a full body and free from the sharp acid taste which colonial wines often have.' The sample of cherry brandy was 'deliciously flavoured liquor, and the orange wine and bitters are of superior quality.'[14]

Figure 9. 5. The winery built by Feraud in 1871 is the only remaining original building on the former 'Monte Christo' property. It has lately been restored, and is in use as a dwelling and office building.

Feraud went on to win a number of First Class merit awards at Sydney and Melbourne International Exhibitions and at the Dunedin Industrial Exhibition during the early 1880s.[15]

'Monte Christo,' after being leased for a few years, was eventually sold in 1886 to William Angus of Dunedin, who sold it in 1889 to James Bodkin of Queenstown. The Bodkin family farmed 'Monte Christo' for the next 66 years when the place began to be subdivided into smaller blocks.[16]

Local Body Politician

When Feraud was elected as the first mayor of the new Clyde municipality in 1866, he was at the height of his popularity with the public. In fact, in March 1867, he was approached by a number of electors to stand for the Dunstan Electorate of the Otago Provincial Council. However, he respectfully declined, saying that his 'various avocations precluded the possibility of accepting.' He was appointed a J. P. at this time.

Feraud was elected unopposed in July 1867 for another term as Mayor but his short fuse and unpredictable temperament was his undoing. Members of the Hospital Committee of the Borough Council[17] would not support him in several staff disciplinary matters, and he resigned from the committee in August 1867. When the matter was taken to the council meeting in October, he handed in his resignation from the mayoralty 'for business reasons.' His letter was held over while the Town Clerk asked him to reconsider, but his resignation was accepted at the November meeting. Shortly afterwards he received a letter of thanks for his services.

Feraud was then elected as a councillor to the Borough Council in June 1868 but he became involved in a row over the siting of the proposed new Town Hall. The choice was between the site of the old hall in the centre of the town or a reserve set aside by the Provincial Government for the purpose, on the outskirts. At the October meeting the council was evenly divided on the matter and the mayor's vote favoured retention of the existing site.[18] The defeated councillors, including Feraud, promptly resigned.

The mayor called a public meeting that seemed merely to provide a forum for the councillors to continue their fight. It was, however, disclosed during unseemly name-calling that a number of councillors had a conflict of interest, in that they wanted the hall near their businesses. This applied particularly to Mayor Hazlett whose place of business was just across the road from the proposed central site for the hall.[19] So convincing was the mayor's rhetoric that the public meeting not only expressed strong backing for the mayor, but also passed a resolution that the council accept Councillor Feraud's resignation. These resolutions strengthened the resolve of the three councillors who had resigned, including Feraud, who then asked that their letters of resignation be

Figure 9.6. The siting of the new Town Hall was the cause of Feraud resigning from Clyde Borough Council. Designed as a Masonic hall, it was taken over as a town hall when the Masons ran out of money. Built in 1868-69 it must have seemed a pretentious building compared with its neighbours of the time.

withdrawn—they decided to remain on the council and fight the mayor and his devious schemes. This led to a farcical situation at the November meeting.

The mayor and his supporters were late in arriving, so the councillors already present voted Feraud to the Chair and commenced business. When the others arrived Feraud immediately vacated the Chair, but the mayor declared the early meeting illegal as a quorum was not present. There was a tussle over possession of the Minute Book but the mayor, James Hazlett, a big man, gained possession of it and locked it away in the safe, at the same time declaring the meeting adjourned. All councillors were notified that the meeting would be reconvened next morning—all that is, except Feraud. When a councillor questioned the legality of holding the reconvened meeting without having notified Feraud, he was told by the mayor that Feraud was no longer a councillor and produced Feraud's letter of resignation of two months before as evidence! The Town Clerk then complained that a councillor had been seen tearing up a letter—presumably the one from Feraud asking that his resignation be withdrawn. The letters of the other two councillors asking for withdrawal of their letters of resignation were accepted, but not that of

Figure 9. 7. James Hazlett was Mayor of Clyde during Feraud's sojourn on the Borough Council. Hazlett was a member of the Otago Provincial Council, and in 1878 joined James MacKerras to form the well-known Dunedin merchant firm, MacKerras and Hazlett.

Feraud. He had been railroaded out of Council. Feraud was not to forget the behaviour of Mayor Hazlett towards him. Council, however, had apparently grown tired of Feraud's tantrums, as according to the newspaper reports:

> Councillor Feraud's resignation was accepted without comment, and the elements of disturbance no longer present themselves in the Council.[20]

This was amplified a week later:

> Councillor Feraud is certainly a very enterprising and energetic man but there are members of the Corporation equally energetic and possessing as much common sense, education and with equal standing in the community. As evidence that the mayor was right and Cr Feraud wrong it may be mentioned that his resignation was accepted without comment by his readiest supporters.[21]

It was about this time that a 'parking warden' in the guise of the Inspector of Nuisances, told Feraud to move his parked buggy. A reporter, who saw the incident, suggested that Feraud might follow the example of the Mayor of Melbourne and get a donkey as it took up less room. Feraud replied that there were already enough donkeys on Council.

The Man

Jean Feraud was a clever, industrious and innovative man who gave generously of his time to the community. He took part in a number of deputations to the Superintendent of the Province in Dunedin, was a foundation member of the District Improvement Society and a member of the Hospital Committee. Feraud's home and garden were made available for community functions and his carriage for the transport of children to picnics. He and his wife were generous with hospitality and graciously entertained Governor Grey during his tour of Central Otago in March 1867. He was a regular donator of fruit and vegetables to fund-raising schemes, particularly those connected with the hospital.

Feraud produced far-sighted, if impractical, schemes. One proposed diverting the Clutha River from near Clyde into the Manuherikia River, thus exposing many miles of, hopefully, gold-bearing river bed. The only difficulty he could envisage was finding a suitable foundation for the necessary dam near Clyde. He advocated a system of locks on the Clutha River to make it navigable for small steamers, as had been done in the south of France.

Figure 9. 8. Air view of part of Dunstan Flat downstream from Clyde. 'Monte Christo,' now much subdivided, lay in the centre of the photograph to the left of Youngs Lane (the prominent road just to the right of centre pointing directly to the mouth of the Waikerikeri Valley. Highway 8 in the foreground.

In a thoughtful letter to the *Dunstan Times*, Feraud advocated the introduction of 'four or five thousand' Chinese to the gold fields to replace the loss of miners to the West Coast gold fields. He pointed out that up-country storekeepers were languishing and owed Dunedin merchants something like a million pounds. Quoting the axiom that 'population is the wealth of nations,' he went on to extol the virtue of Chinese, pointing out that they were painstaking, industrious and energetic. At the same time Victorian experience showed they were peace loving and settled down to become excellent gardeners and cultivators of the soil. He was ready, he said, to donate £20 towards securing 'the introduction of a large body of such a patient, industrious and useful class as the Chinese gold diggers.'

This letter appeared more than a year before the first Chinese diggers arrived at Clyde.[22]

There is no doubt that Feraud was an impatient, short-tempered man who had difficulty in getting on with people. Perhaps it was his inability to discuss matters rationally and make compromises that led him so often to take his fellow citizens to court on matters ranging from damage by wandering bullocks, to flooding of his property. However, it was only when he began his long drawn-out legal battle with James Holt over the water of Waikerikeri Creek and the people of Clyde perceived a threat to the town water supply, that public sympathy turned against Feraud. Details of this battle are set out in the next chapter.

The Family

Feraud married an Irish lady, Anna Maria Connor (?), who was some years older than him. A daughter, Frances, was born in 1855, but presumably died before they came to New Zealand in 1863. They had another child at Clyde, a son, Jean Louis, who died aged two, in 1866. It is said that Feraud himself suffered from long bouts of illness. The family left the Dunstan in 1882 and lived and worked in Dunedin until 1888, when they moved to Adelaide where Feraud once more set up a wine and cordial business. Feraud died there in 1898 aged 78 from 'senile decay.' He is buried in the Catholic section of West Terrace Cemetery, Adelaide, but owing to a fire in 1904 that destroyed the cemetery records, his grave has not yet been located. He was known as 'John Ferand,'[23]

Mrs Feraud lived on, but met a tragic end 18 months later:

> An old lady, named Ann Marie Serand (sic), aged 86, widow, residing in Chapel Street, Norwood, was burned to death in her home on Saturday. She was the sole occupant of a two-roomed detached cottage, and was last seen at about 2 o'clock on Saturday afternoon. . . At half past four a neighbour who was passing detected the smell of fire, and with difficulty the back door was pushed open, and the charred body of the old lady was found on the floor. A broken kerosine lamp was lying close by. . . The Coroner decided not to hold an inquest.[24]

The Bladiers, with their Australian experience, were almost certainly the first to introduce the practice of irrigation to Central Otago but they were not able to demonstrate its full benefits before they left the district. Without doubt, though, they had passed their knowledge on to their former friend and neighbour, Feraud, who applied it in his development of 'Monte Christo.'

Jean Desirè Feraud, a colourful personality, deserves a place in the history of Central Otago on several counts. First, he amply demonstrated, by the productivity of his farm, the inherent fertility of Central Otago soils if they were provided with adequate water. He applied the methods of irrigation (probably furrow irrigation) first introduced by Bladier. Secondly, he was the first to successfully produce good quality wines in commercial quantities, and is justly regarded as the father of the wine industry which, though forgotten for nearly 100 years, is undergoing a dramatic resurgence today.

Notes

1. Information about Feraud's early years kindly supplied by Wayne Stark of Christchurch who is researching the life of Feraud and his friends, the brothers Bladier.
2. Annual report of the subscribers to the Dunstan District Hospital *Otago Daily Times* 31 January 1865.
3. *Otago Daily Times* 14 June 1864. In many reports Feraud' is spelt 'Ferand.' This can be attributed to the fact that most writing, even at the turn of the century, was in longhand. In most longhand writing 'u' and 'n' can be very similar.
4. For description of the Frenchmans Point claims see McCraw, 2000.
5. It is not certain whether both brothers or only Jacques was involved in this enterprise.
6. Records of Clyde Wardens Court held in Dunedin Office of National Archives.
7. *Otago Provincial Gazette* No 275, 11 November 1863. pp. 480-481.
8. Within the block, but not included in the total acreage, was a tailrace of 4 acres.
9. Mrs Bladier and 3 children arrived at Port Chalmers in 1863. It is not known whether they moved up to the Dunstan. In August 1866 the whole family landed in Hokitika from the *Gothenburg* so it is presumed that they had returned to Australia in 1865. It is thought that they took up farming. In 1869 another Mrs Bladier (perhaps the wife of a Bladier brother) arrived in Hokitika with two children.
10. Feraud may well have had an interest in the Monte Christo gold mine near Ballarat in Victoria. 'Monte Christo' is well described in Veitch, (1976) and in Bodkin (1998).
11. *Dunstan Times* 26 July 1867.
12. *Otago Daily Times* 27 February 1869.
13. *Tuapeka Times* 1 April 1872.
14. *Tuapeka Times* 1 April 1872

15. Betty Veitch 1976 p. 60.
16. A. W. Bodkin 1998 pp. 14-15 and 86-87.
17. 'Borough Council' and 'Town Council' was used synonymously. There were also references to the 'Corporation of Clyde.' or just 'Corporation.' The Municipality was regarded as an incorporation of Mayor, Council and Citizens.
18. *Otago Daily Times* 23 October 1868.
19. *Otago Daily Times* 2 November 1868.
20. *Otago Daily Times* 19 November 1868.
21. *Otago Daily Times* 27 November 1868.
22. *Otago Daily Times 23* March 1865.
23. Information kindly supplied by Robert Blair of the Genealogical Society of South Australia.
24. *Register* (Adelaide) 20 November 1899.

10.

THE FIGHTING FRENCHMAN

—Feraud's Battle for Water

At a public meeting in 1864 called to discuss a threat to the town's water supply, the citizens of Clyde were ambivalent. On the one hand this man Feraud was threatening to cut a water race out of Waikerikeri Creek to supply his extensive gardens. Such a race would certainly interfere with the water available for Holt's race, which supplied the town, and might leave the citizens short, or even without water. On the other hand the people of Clyde and the surrounding district depended on Feraud's gardens for fresh vegetables and fruit.

Holt's Race

Holt's race supplied water not only for the domestic use and fire protection of Clyde but also water to drive the pumps at the coal mine. Without coal Clyde would have been uninhabitable during the long, bitterly cold winters.

James Holt had arrived with the Dunstan Rush, and no doubt took part in the early mining frenzy along the banks of the Clutha River before the river rose in September 1862 and swamped the claims. Instead of taking to the hills as a prospector, Holt opted to become a retailer of water. He and his mate Roscoe were granted a licence for a nine mile (15 km) race from Waikerikeri Gorge to Muttontown Point, and a water right of two heads from Waikerikeri Creek, under the usual conditions that the water was to be used for gold mining.

At Muttontown Point a number of miners were working on the flood plain of the river, but in spite of the fact that the largest river in New Zealand was flowing past a few metres away, they were desperately short of water. The reason was that the river fluctuated in level from day to day, and it had proved almost impossible to find a way of feeding water from it into a water race. Holt reasoned that there would be a ready sale for water at Muttontown Point.

Figure 10. 1. William Holt and wife Kitty. For 14 years Feraud fought Holt in the courts for a share in the water from Waikerikeri Creek.

During the early part of 1863 (long before the Agricultural Reserve was proclaimed) Holt and his men had begun cutting a race from the mouth of the gorge, down the flat floor of the valley to the point where the valley leaves the high terrace.

At some stage Holt changed his mind about taking his race down to Muttontown Point. It is likely that it was a commercial decision. Morgan had opened up a coal mine on the riverbank at Clyde and needed water to drive his pumps, and no doubt the citizens of Clyde were keen to have a water race to supply their domestic requirements. Whatever the reasons,

Holt decided to divert his race to Clyde,making provision to supply a small amount of water to the hospital on the way.

This diversion of Holt's race was no doubt a disappointment to the miners of Muttontown Point who were expecting water. So one of their number, Andrew Dalziel, began to construct a race from Muttontown Point to an intake in the lower part of Waikerikeri Creek where, apparently by prior arrangement, he expected to receive an allocation of water from Holt's race.

Dalziel's applications to the Warden's Court tell the story. First, in January 1864 he asked for an intake in Waikerikeri Creek 'near Bladiers' garden' which was planned to pick up water he was to purchase from James Holt. But Holt, committed to supplying the coalmine at Clyde, could not or would not, now let him have any water. A few months later, in a fresh application to the court, Dalziel asked for the intake to be moved above the 'Dunedin road' (the present Springvale Road) because he 'could not get a head of water from Holt.' Presumably Dalziel now intended to draw water from the spring-fed Waipuna Creek, but he found the supply insufficient, because in August 1864 another application was made for a race nine miles long from the gorge of Waikerikeri Creek to Muttontown Point. Dalziel was granted a head of water second in priority to Holt.

Although no map of Dalziel's race exists, it is assumed that it was cut along the eastern side of Waikerikeri Valley because it picked up the water from the spring-fed Waipuna Creek on the way.

Now Holt and his new race ran into a difficulty. The Provincial Government had proclaimed, in November, that several towns, including Clyde, were to be exempted from the operation of the Goldfields' Regulations.[1] This meant no mining in the towns.

At the end of the year Holt's water race licence came up for renewal. It had been granted, like all licences under the Act, for mining purposes, but now mining in Clyde had been prohibited. This presented the warden with a dilemma. Strictly, he should have not have renewed Holt's licence, as his licence was for mining. But here was a water race, which had cost a lot of money, ready to be used; a coal mine desperately needing the water to operate; citizens half frozen for lack of coal and clamouring for a supply of water for household purposes.

The warden gave in. He renewed Holt's licence endorsed with the words 'For domestic purposes of the Town of Clyde,' which apparently was also to cover use in the coal mine. A further condition, suggested by the Superintendent, was that water had to be made available for the future agriculturists in the Waikerikeri Valley. Although it seemed a commonsense solution, the warden had acted outside the law of the time. Clyde's water supply was illegal.

So far, neither the copy of the original licence issued to Holt, nor the letter sent by the Superintendent to the warden in 1863, have been located amongst the voluminous records of the Dunstan Warden's Court. The letter pointed out that the Government intended to proclaim an Agricultural Reserve that would include the Waikerikeri Valley, and went on to suggest that Waikerikeri Creek water should be for the benefit of future settlers in the valley. We have only the newspaper account of Holt's version of what was an unusual case.

Feraud held that in an Agricultural Reserve the settlers had the right to use the water of streams flowing through the reserve, and his opinion was supported by a statement by the Secretary of the Goldfields, Vincent Pyke. Although Holt had agreed, as a condition of his licence, that some water would be allowed to flow down the creek for the farmers, there never seemed to be any water in the bed of the lower reaches of Waikerikeri Creek. After Holt's race gathered its share, the remainder simply soaked away into the gravel bed of the stream.

Feraud and the Bladiers approached Holt a number of times to try to make some arrangement to draw water from his race, but without success. They must have found it galling to have water, to which they were sure they were entitled, flowing within a few yards of their boundaries, and yet not have access to it.

Trials of a Pioneer Irrigator

Feraud was almost certainly correct in his opinion as to who could use the water, but as it turned out, there were differences of opinion as to what 'use' meant. When their overtures to Holt were rejected, Feraud and Bladier, clearly exasperated, threatened to cut a race of their own from the gorge of Waikerikeri Creek. This so alarmed the citizens of Clyde that a public meeting was called to discuss the threat.[2]

At this meeting Feraud made two statements which clearly guided the course of his actions over the next fifteen years. First, he said that he had understood that the water in Waikerikeri Creek was for the use of occupants of the Agricultural Reserve, and he had taken up his land on that understanding. Secondly, he made it clear that it was not his intention to deprive the townspeople of their water.

During the course of the meeting Feraud suggested that Holt, who had recently taken over the coal mine, could equally well use a steam engine to drive his mine pumps and so free up the water for others. Holt countered by saying he had already spent £600 on mine machinery and could not afford to change his power source. Holt went on to say that he thought Mr Bladier used water very extravagantly on his open, porous ground and no doubt would use every drop that was in the race, and more, if he were given the chance.

The large crowd at the meeting was confused. They did not want to deprive the farmers of water entirely, but could not have them diverting all the water from Waikerikeri Creek, as Feraud was threatening to do. One speaker, a well-known citizen, rightly said that if the coal mine stopped work, half the population of the district would leave. He then went on to say that he had never heard of farmers diverting the water from creeks into water races. Surely Mr Pyke, Secretary of the Goldfields, had not envisaged such a thing when he said the water in the creek was

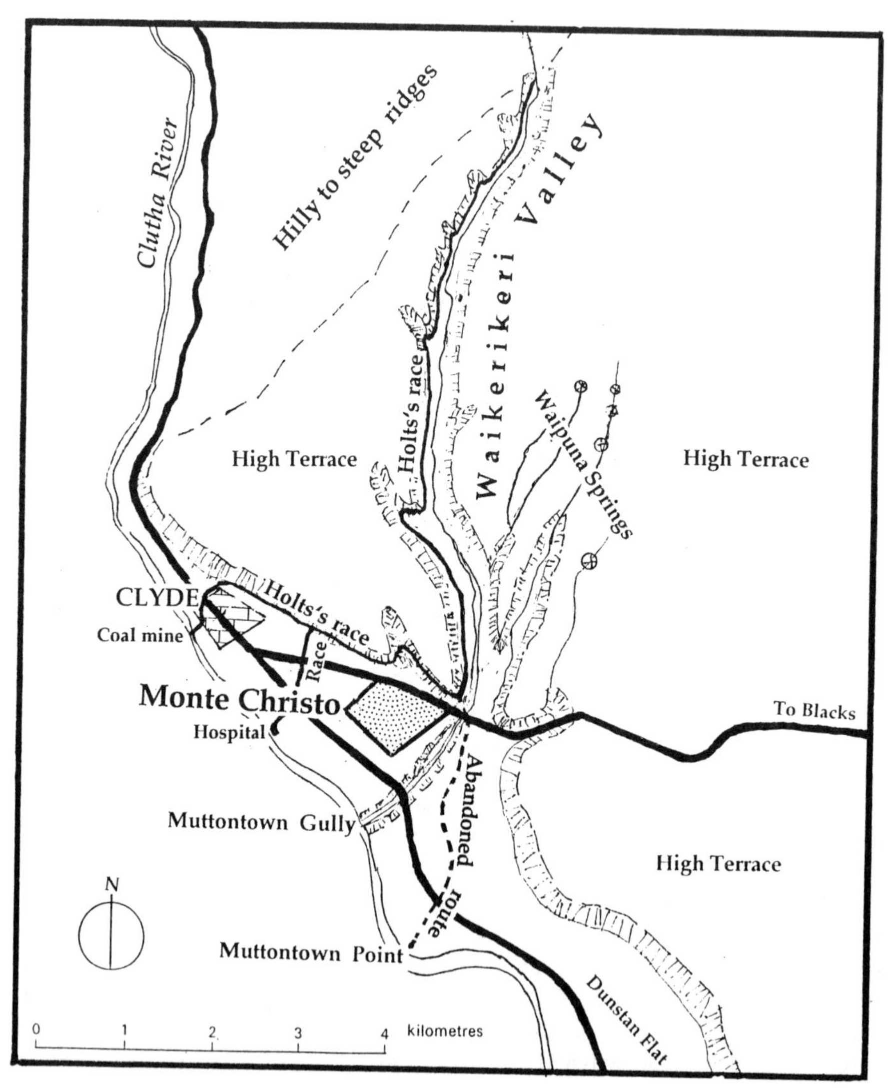

Figure 10. 2. Holt's race was planned to supply water to miners at Muttontown Point, but instead it supplied water to the Clyde coal mine and the township, passing 'Monte Christo' on the way. A branch race supplied water to the hospital.

for the use of agriculturists. It was suggested that the farmers could obtain a good supply of water from bores sunk to a depth of 35 ft. In the end the meeting drew up a petition, which was eventually signed by almost everyone, and forwarded to Vincent Pyke. In summary, it asked that the respective rights of the agriculturists and the townspeople, so far as water supply was concerned, be clarified and settled once and for all.

This meeting was a milestone. Obviously it was the first time people had heard about 'irrigation,' as distinct from 'watering.' It was clear they had no idea of what the irrigation of a large orchard/market garden involved. Diverting water from creeks into water races was something miners did, but not farmers. Farmers watered their plants and trees from buckets of water drawn from a well or stream. Why couldn't Feraud do the same? If nothing else, the meeting provided proof that nothing was really known of irrigation before Bladier started his garden, and this helps to confirm his right to be regarded as the pioneer of irrigation.

A month later the petitioners had a reply from the Superintendent of the Province, Mr Hyde Harris, which satisfied them in the short term, but did not bode well for the future. The Superintendent pointed out again that Holt and his partner Roscoe, had themselves proposed, as a condition upon which the water right should be issued to them, 'that the agricultural tenants in the Waikerikeri Valley should always have an ample supply of water.' The Superintendent pointed out, however, that Bladier and Feraud's land was not actually in the Waikerikeri Valley, but was on Dunstan Flat and therefore they were not, as of right, entitled to divert the Waikerikeri water to their properties.

This reply gave so much 'universal satisfaction' to the citizens of Clyde that they didn't take too much notice of the remainder of the same letter. It went on to warn that the floor of the Waikerikeri Valley had already been surveyed, and the land would shortly be offered for sale or lease. When these freeholders or lessees applied for water, 'the water now diverted to Clyde must be returned to its natural river bed.'[3] Feraud, however, took notice.

The Superintendent's reply showed that he, too, did not understand irrigation. It seems that when it was suggested that the waters of Waikerikeri Creek were for the use of the agriculturalists, everyone had in mind farmers settling along the banks of the creek, their stock drinking from the stream and the farmers carrying their domestic water up to the house in buckets. They would be, in effect, exercising their riparian rights. Hence the Superintendent's point that Feraud, although a farmer, had settled away from the stream so had no claim on the water of Waikerikeri Creek. The Superintendent foresaw the necessity of returning some at least, of the Clyde town supply back into the creek, so that the farmers who were going to settle along the valley could draw stock and domestic water from the creek. The Superintendent had not thought of anyone wanting to transport water in quantities to a garden and orchard by means of a water race. And neither, obviously, had Holt when he generously promised 'ample' water for the farmers along the stream.

In spite of their initial lack of success in obtaining a plentiful and reliable supply of water from Waikerikeri Creek, Bladier and Feraud pressed on with the development of their farms. They drew what water they could from the Waipuna Springs Creek, a tributary of Waikerikeri Creek, but they were so convinced that they had a right to the water from the creek itself that they were prepared to try 'direct action.'

Andrew Dalziel owned a race that drew one head of water, when it was available, from Waikerikeri Creek just above the intake of Holt's race. His was 'second right,' which meant that Holt, who held the first right, got his water before Dalziel could have his. Holt was in the habit of taking almost all of the water in the creek, but Bladier and Feraud persuaded Dalziel to let them use his race for a time. They first made sure that Holt got only the bare minimum water he was entitled to, and they then diverted the rest into Dalziel's race and thence into their gardens. With only the bare two heads entering Holt's poorly maintained and leaky race, practically none arrived at his mine in Clyde and he had to close down for three days.

After warning Dalziel to stop diverting his water, Holt took him to court for £12 damages. The Warden listened to Bladier and Feraud claiming that as agriculturalists they were entitled to the water, but he pointed out that the right of the agriculturists had not been clearly defined. He was doubtful if it were the role of the court to attempt to do so. Holt had on a previous occasion been told that he had to put his race in order to the satisfaction of the warden, but as he had not done so, he had no right to sue. In effect it was Holt's own fault that no water got to the end of his race. He lost the case. The warden said he would refer the vexed question of the Waikerikeri Creek water to the Superintendent for a final settlement.[4]

Feraud remembered the Superintendent's 1864 warning to the petitioners — when the Waikerikeri Valley land was finally taken up by settlers and they applied for water, 'the water now diverted to Clyde must be returned to its natural river bed.' With foresight, Feraud began applying in late 1865 for leases over the flat land on the floor of the Waikerikeri Valley,and by 1870 had taken over about 200 acres (80h). The leased land began about one kilometre up from the present Springvale Road crossing, and it took the full width of the valley floor and stream bed, for a distance of five kilometres up the valley. This Waikerikeri Farm, as Feraud called it, was used mainly for cropping, but in the upper valley where there was some moisture available from seepage from tributary valleys, there was sufficient grass to sustain some dairying.

Now as the major landholder in the valley, and with land bordering the Waikerikeri Creek for a long distance, Feraud had greatly strengthened his right to claim the water in the creek.

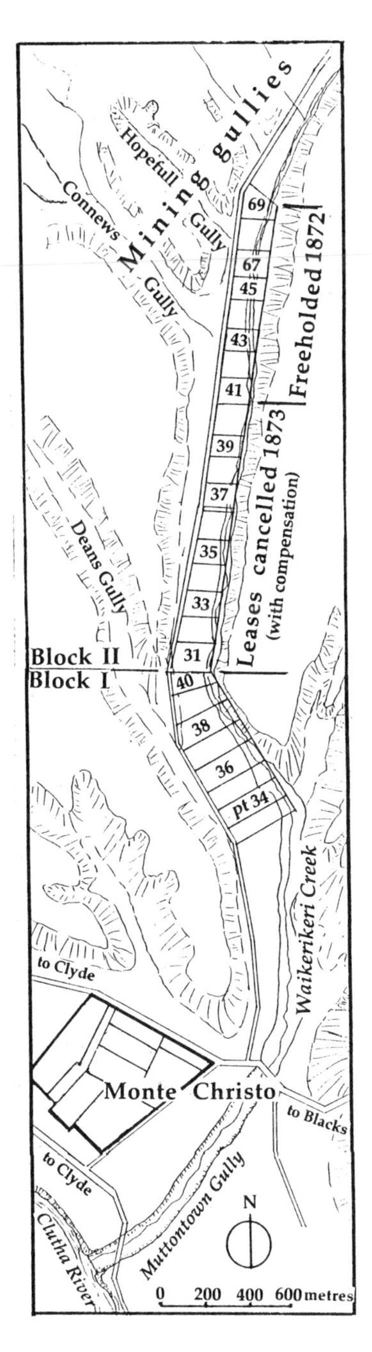

Figure 10. 3. Map showing 'Monte Christo' and the 200 acres (80 h) leased in the Waikerikeri Valley which comprised Sections 33 to 69 (only every second section numbered on map). It also shows the sections freeholded in 1872 and those given back to the Crown (with compensation for improvements) in 1873.

Fight for Water Begins

By 1870 'Monte Christo' needed still more water for its flourishing and expanding crops, and the leasehold lands in Waikerikeri Valley were mainly without water. Under the Gold Fields Act of 1866, anyone who had leased Crown Land for three years could freehold it for £1 per acre. So in 1868 Feraud had sent in an application to the Waste Lands Board to freehold a large block of his leasehold land in the Waikerikeri Valley. Four years were to pass before he heard the result!

Feraud had already begun a battle for water which was to last nearly 10 years. He had again approached James Holt for sufficient water for four week's irrigation. As a settler on the Agricultural Reserve, and now with land bordering Waikerikeri Creek, there is little doubt that Feraud felt he was entitled to the water as of right. Holt said yes, he could have some water if and when Holt could spare it, but with no guarantee of supply. Feraud wasn't satisfied with this offer and took it as a refusal to supply the water he required.

The opening shot in the battle for water was fired in December 1870, when Feraud took Holt to the Warden's Court to claim £20 a day for damage to his crops caused by the non-supply of water. The warden said Feraud should have taken the offered water, and dismissed the case, but made two important Orders. One was that one head of water should at all times be supplied to Clyde for domestic use and to drive the coal mine machinery. This was equivalent to granting the town a water right. The other Order was that Feraud was granted the use of any surplus water for 8 days a fortnight over the period 15 September to 31 December. He immediately obtained written permission to extend this period until 16 January.

Feraud had only begun to irrigate, however, when the water was cut off. His trees and crops suffered damage from drought.

An angry Feraud then launched a two-pronged attack on Holt. In the Warden's Court during February 1871, he first applied to have Holt's water right forfeited, and then he sued Holt for £200 for damage to his crops. Feraud's case was that Holt was not using the water for mining, and he was not allowing water to pass down Waikerikeri Creek for the use of farmers. In addition, the arrangement Holt had with the Clyde Borough Council, which had bought a head of water from him, was probably illegal.

If Feraud's application were successful, and he gained Holt's water right, it could leave Clyde without water, as it was unlikely that Holt would allow water other than his own, to flow along a race he owned. Feraud had said, in public a number of times however, that it was not his intention to deprive Clyde of its water supply. So he, along with other neighbouring farmers, applied to construct a new race from Waikerikeri Creek to Clyde, where it would supply the town with water. It would then carry on to the banks of the Clutha River where the water could be used for sluicing. This would make the use of the water legal, but would give Feraud and his party control of the water. The arrangement was not as one-sided as it might seem— Feraud would use the water in summer for irrigation, and the miners in the winter when the river was low.

When the people of Clyde heard about this scheme, they again became concerned about the future of their water supply. They had, at the moment, a perfectly good arrangement, especially since December 1870 when the Warden had ordered that one head of water had to be set aside for the use of Clyde township and the coal mine. What would happen if Holt lost his water right and it was transferred to Feraud? Could they trust Feraud, the man who a few years before was lauded with respect and praised for his integrity, but now seemed to many to be determined to deprive Clyde of its water supply for his own purposes? A public meeting decided to formally object to Feraud's application, and there and then, began to collect funds to meet the cost of the Objection.

Error of Judgement

Normally, applications for 'mining privileges' were heard at the following fortnightly session of the Warden's Court. It was the practice though, to adjourn all contested applications to allow the parties to prepare their cases. Sometimes, for various reasons, there was more than one adjournment. When Feraud's application was adjourned, he was suspicious and sensed some sort of conspiracy. Against the earnest advice of his lawyer, and with what the newspaper described[5] as 'cool impertinence,' Feraud sent a Petition to the Provincial Government to the effect that the warden, Vincent Pyke, was denying him justice. The effect on Pyke, a man almost as short-tempered as Feraud, can be imagined. He was furious and had his revenge by insisting that the case be heard in meticulous detail, with many adjournments.

Feraud's case (his co-applicants had pulled out) was familiar. It was that Holt's water right, granted in December 1862, contained a clause which said the water had to be used for mining purposes, and he had not used it for such. Apparently, use in a coal mine didn't count as 'mining' and in fact, Holt admitted that the water was not used for mining. Feraud's proposed race, on the other hand, commanded very rich ground and would be used for sluicing — that is, after Feraud had taken what he needed for irrigation of his orchard, and the settlers along the line of the race had taken their share.

It was June 1871 before the long proceedings came to an end, and on the issue of compensation for damage to crops, Feraud was non-suited on the grounds of insufficient evidence. Each party was ordered to pay his own costs. The application for cancellation of Holt's water right was refused. So the *status quo* was maintained, but Feraud appealed to the District Court.

Judge Gray was more sympathetic to Feraud. He found that, provided Holt took only what he was strictly entitled to, the settlers in the valley were entitled to the remainder. As Feraud was by now the main land holder in the valley, the water was his, albeit only any surplus left over after Holt's race had taken its quota. As an extra, Judge Gray awarded Feraud £30 for the damages to his crops.

So after seven years of battling, Feraud had at last won the right to draw some water from Waikerikeri Creek. And Clyde township still had its water supply.

Figure 10. 4. Waikerikeri Valley. 'Monte Christo' and Bladiers' properties left foreground. Dunstan Road right foreground. Waikerikeri diggings were in gullies in the terraces sloping from the foothills. Dunstan Mountains in the background.

Battle of Waikerikeri Valley

Feraud had begun to lease, from an early date, land on the flat floor of Waikerikeri Valley, and a number of consequences arose from this. First, since Waikerikeri Creek ran through the length of the block, Feraud had access to any water that might be in the stream. Secondly, any miners working on the adjacent hillsides could not discharge their tailings into the Waikerikeri stream bed without crossing Feraud's land. Even though the land had been declared an Agricultural Reserve, it was still Crown land and so could be used for mining, provided it was not occupied by farmers. And even if it were, it could still be mined, provided that compensation for any improvements was paid.

None of this was a problem in the early years of Feraud's leaseholding, but the discovery of gold at shallow depths in the gullies draining down into the Waikerikeri Creek during the early 1870s drastically changed the situation. Miners began constructing large water races, which were to bring in something like 25 heads of water, with the express intent of sluicing these gullies. The necessity for tailraces to get rid of the huge amounts of tailings resulting from mining on this scale was obvious. But if the miners discharged tailings on to Feraud's land they became liable for damages to his improvements.

Just how sensitive Feraud was about damage to his land had been demonstrated in December 1869 when he laid a 'plaint' against Holt for damage to his fences allegedly caused by water escaping from Holt's race.[6] The matter went to arbitration and it turned out that a severe rainstorm, a 'cloudburst' in fact, had swept down the gullies, with water five feet deep according to one witness, filling up Holt's race and spreading gravel and sand over Feraud's leasehold. The arbitrators had no problem fixing the blame on an Act of God and made Feraud pay all the expenses of the investigation.

It was matter enough for concern that Feraud's leasehold land lay between the intended mines and the repository for their tailings, but news had just arrived that the situation had suddenly become much worse. The Waste Lands Board had at its meeting of 11 December 1872, sanctioned the sale of eight sections, including some required for tailings disposal, to Mr Feraud.[7]

The people of Clyde could not believe it. Miners would now require Feraud's permission to discharge their tailings, and even if he granted this, which was highly unlikely, they would have to pay compensation for any damage not only to his improvements, but also to his land. The future of mining at the Waikerikeri Diggings had been looking particularly bright with two large water races about to be brought on to the ground, but without a place to dump tailings, mining would be impossible. When the miners and the public realised the implications of what was happening, there was an outcry.

An urgent public meeting was called. Everyone at the meeting knew, or was soon informed, that as long ago as 1869 the Provincial Council had agreed that only in unusual and exceptional circumstances should the Government allow leased goldfield's land to become freehold.

Furthermore, only six months before, a Petition signed by 250 'Merchants and Others of Dunstan' had been presented to the Provincial Council by the Member for the Goldfields, Mr Shepherd. Shepherd was also Chairman of the Goldfields Committee of the Provincial Council,

Figure 10.5. Upper Waikerikeri Valley from the outlet of Deans Gully. All of the flat valley floor in the view was leased by Feraud. Gold was mined in gullies in the sloping terraces along the foot of the range. Dunstan Mountains in the background with Leaning Rock prominent on the skyline

whose task was to consider the petition and make recommendations to Council. The Committee's recommendations[8] had been clear:

1. . . . all lands in the Leaning Rock Survey District, Dunstan, alleged to be auriferous by the petitioners, be reserved from leasing . '
2. That all auriferous land in the Leaning Rock Survey District already leased should not be sold to the lessees.

The petitioners had been notified that the Leaning Rock block had been withdrawn from lease or sale.

Now telegrams were flying backwards and forwards between Clyde and Mr Shepherd, and they were read out to the meeting. It seemed that it was all too late. The sale had been completed a fortnight before, and according to Mr Bastings, Secretary of the Goldfields, nothing could now be done. Shepherd was embarrassed, and blamed the Executive and Administration:

> [My recommendations have been] entirely ignored by the late and present Executives. The position of the goldfields is an unhappy one, handed over from one ignorant Executive to another. I have done everything I could, and it is most disheartening to be frustrated by executive maladministration.[9]

Hard words were said at the public meeting. The Government had 'ignored the wishes of the mining community,' 'treating us worse than Chinamen' (whatever that might have meant), and so on. Finally a deputation of two was appointed to go to Dunedin to sort things out, and it was resolved 'That this meeting has no confidence in the Executive as at present constituted' And with a hearty vote of thanks to the Chair the meeting was terminated.

The deputation met the Superintendent and Bastings who did all the talking. He said that the present Government knew nothing of the sale and blamed the Waste Lands Board. He said endeavours would be made to come to terms with Mr Feraud about getting back the land that had been purchased. He assured the deputation that the land would not then be resold.[10]

Feraud had done nothing illegal. Clause 52 of the 1866 Gold Fields Act said clearly that anyone who was the holder of leased land on a goldfield could purchase it after three years, providing that they had cultivated, or otherwise improved, two thirds of the area. But how could a responsible Provincial Government, through its Waste Lands Board, overlook promises made to the Clyde people and ignore recommendations, which it had approved, from its Goldfields Committee? Was it concerned with having to pay compensation if it had refused the sale? Surely not!

Mr F. W. Wilson, solicitor, of Clyde, took it upon himself to investigate the reasons. He found that Feraud had applied to purchase the land four years before, in December 1868 — years before there was any talk of gold, water races or tailings in the Waikerikeri Valley. The Government approved the sale, but because an Education Reserve had been included in the lease by mistake, the matter was laid aside for further investigation, and then apparently forgotten. Only recently had the Education Reserve come up for consideration, and been cancelled. Feraud's four year old application was found, the original Minute of 1869 approving the sale still attached to it, so the whole thing was sent to the Waste Lands Board who had no option but to grant an application approved by the Government. Wilson blamed the Secretary of the Goldfields who should have known about the Government's decision on the petition.

Much of the business of that January meeting of the Waste Lands Board was taken up with the Feraud affair. In spite of Basting's telegram (he was also Chairman of the Waste Lands Board) to the public meeting in Clyde only a week before, saying it was impossible to do anything about the sale, the Board decided to hold a re-hearing.

Feraud also sent a letter to the Waste Lands Board about the eight leasehold sections he held immediately down-valley from the ones he had been allowed to purchase. He demanded, in terms of the Act,[11] that the Board cancel the leases on these sections and pay him for improvements.

He had applied to the Waste Lands Board to freehold them also, but the Board had applied the new rule and turned him down.

Feraud was short of cash and was, in fact, declared bankrupt a few months later so perhaps did not want to, or could not, continue paying rent for the sections.[12] On the other hand he may have deliberately applied for the freehold knowing that under the new rules he would be turned down and this would leave the way open for him to claim compensation. Determining the amount of compensation for cancelling the leases required arbitration, and the sum of £220 (about $NZ35,000) was agreed on at a hearing held in May.

The Superintendent Visits

The Superintendent of the Province, James Macandrew, decided that February 1873 would be a suitable time to make a tour of the goldfields. He was accompanied by Bastings, the Secretary of the Goldfields, as well as various Provincial Council members and officials. At Clyde the entourage was joined for a local tour by the Mayor of Clyde and his full Borough Council, the Warden and 'a number of other gentlemen,' including newspaper correspondents. They were entertained by Feraud at 'Monte Christo' before setting off on horseback. Local miners explained the tailings disposal problem, and pointed out that, on this gently sloping land, a simple channel was not sufficient as it would simply block up—an expanse of land was needed so the tailings could spread out across the flood plain. Feraud didn't like the way the conversation was going:

> Some little recrimination here took place. Mr Feraud protesting that Monte Christo was of no use without the block up the valley, and some words the reverse of complimentary passed between him and several other gentlemen much to the amusement of the thirty or forty diggers who were there.[13]

Meanwhile no date had been set for the re-hearing of the decision to sell the leased sections and although newspaper correspondents wondered when it was going to be held, it became obvious as time went on that there was not going to be a re-hearing.

In fact, it was not until Feraud petitioned Parliament in 1880 complaining about damage to his freehold land by mining operations and asking Government to purchase the land, that it was decided to resume the land and pay Feraud £300 for it.

It is probable that it remained Crown Land until Mrs Ellen Athfield acquired the block in December 1949. She was issued with a 'Certificate in lieu of Grant' which states the block 'to have been originally acquired by Ellen Atfield.'

Back to the Water

On appeal in July 1871, Feraud had won the right to draw some water from Waikerikeri Creek, but in practice it didn't mean much. Holt had the prior right for two heads and for much of the year, including the irrigation season, Waikerikeri Creek was dry below his intake. But if Feraud could store water from the spring thaw until the irrigation season it would solve most of his problems.

On 3 September 1873 Feraud went before a Waste Lands Board of Inquiry to ask permission to erect a storage dam on Crown Land in the upper Waikerikeri Valley, and construct a water race from it to 'Monte Christo.' Feraud was not happy about the composition of the panel conducting the Inquiry. In fact he had written twice to the Waste Lands Board objecting to Mr Hazlett's presence but the Board had not sustained his objection.[14]

At the Inquiry[15] there were submissions from the miners who asked that nothing be done that would interfere with mining. The citizens of Clyde submitted that after they had taken the water that they were entitled to from Waikerikeri Creek, there was none left in the creek. Feraud's submission was well prepared. He began with the well-canvassed line that the conditional water right held by the Clyde Corporation was not legitimate, as the water was not being used for the purpose for which the right was granted. However, the Inquiry pointed out that it was not sitting in a judicial capacity and would not get involved in arguments about whether rights were legitimate or not. The Rights were there and the Board must recognise them. They had been granted by the Crown and must remain until they were cancelled.

Feraud assured the Inquiry that he could not interfere with the town water right, as his proposed intake was a mile below the intake of Holt's race, and nothing he was doing would interfere with mining. In fact he promised that he would take the responsibility of protecting his works against damage by mining, and would also undertake to give up the land, without compensation, if it were required for mining. Some of the audience thought that this was too good to be true and wondered aloud how, if the application were approved, such conditions could be enforced legally.

The Inquiry rejected the application on the ground that, if granted, it would create the expectation of a right that simply could not be supplied. It had been shown that even if the whole of the Clyde Borough water right were allowed to flow down the creek, the gravelly bed would absorb it before it got to the proposed dam. On hearing this decision, Feraud exploded with anger and frustration and was heard to mutter that he would have Holt's right cancelled within a month.

A letter to the paper written just after the Inquiry[15] reveals some interesting undercurrents. Six months before, it said, Feraud had first written to the Waste Lands Board to object to Mr James Hazlett sitting as a member of the proposed Inquiry panel. Mr Hazlett was a member of the Provincial Council and of the Clyde Borough Council, and therefore an interested party who could not be regarded as unprejudiced. It should not be forgotten that he was also the mayor who forced Feraud to resign from the Borough Council.

When the Board over-ruled the objection, Feraud relied on Mr Hazlett 'to do the right thing,' but Hazlett chose not to relinquish his seat on the Inquiry. The letter to the paper suggested that Hazlett was interested in maintaining Holt's race through his paddocks to provide water for his Shetland ponies which were his pride and joy. It was also revealed that the real objection to the application was the fear that once the dam and race had been approved, Feraud would use his rights as a land owner bordering the creek to demand that Holt's and the Clyde water be returned to the natural channel to be picked up in his race. Finally the writer pointed out that the argument put forward, that returning the water of Waikerikeri Creek to its natural channel would deprive Clyde of its water, was fallacious. The Borough Council had been given power to acquire water by a special Ordinance[16] passed by the Provincial Council only two months before, and had already bought Hastie's water rights and races. This was Dalziel's old race which had a water right for one head and had once drawn this water from the gorge of Waikerikeri Creek above the intake of Holt's race. However, as the water right had only second priority to Holt's, the race also drew more reliable water from the Waipuna Springs located in a tributary valley to the Waikerikeri Valley.

Clyde Borough Council had bought the Waipuna race, as it was called,

Figure 10. 6. Air view of the tributaries of Waipuna Creek. Springs are marked by trees.

as insurance against Feraud making good his threat to capture Holt's water, and also to provide extra water when the flow in Waikerikeri Creek was low. If necessary, Clyde could survive without Holt's race water.

Ten years before, the Superintendent had warned that landholders would demand the return of the diverted water to the creek, and now it had come to pass. In the Warden's Court in June 1874, Feraud applied for a Certificate of Forfeiture against Holt's race. The same arguments were gone over again about illegal use of water.

Taking into account the public use to which the water was being put, the warden found forfeiture was inappropriate. Instead Holt was fined, (on what grounds is no longer clear) and the fine amounted to only one shilling more than the costs of the action.[17]

Feraud felt he had lost the case and appealed the decision. Judge Gray took a long time to make up his mind and, in fact, he didn't. Instead he talked of preparing a case for the Supreme Court which could decide the status of a race that ended in land (Clyde) which was excluded from the Goldfields' Regulations. Eventually it was apparently decided that this course would take too long, and be too expensive (Feraud had been declared bankrupt the year before) as there is no record of the matter ever reaching the Supreme Court.

For four years Feraud was relatively quiet. He may well have been suffering from his recurrent illness, but he was able to produce wines that would become medal winners in New Zealand and Australia. Then on 11 January 1878, with renewed energy and finances, he applied to the Warden's Court to have the water licence of the Corporation of Clyde cancelled, as the race had not been used for a long time.

The Waipuna Race

In spite of what Gilkison[18] and local stories might imply, it is clear that Feraud's latest attack was not against Holt's race, which was Clyde's water supply. It was against the Waipuna race, which had been bought from Hastie as a back-up supply in case Feraud succeeded in his perceived plan to capture Holt's water right. The Waipuna race had been cut down the eastern side of Waikerikeri Creek in 1864 by Andrew Dalziel's party to carry a head of water from Holt's race to the claims at Muttontown Point, but the arrangement with Holt had fallen through. Dalziel had then extended his race north to take in water from the Waipuna springs, and eventually right up to the gorge of Waikerikeri Creek after he had been granted a second right of one head. The race was sold to Hastie and then acquired by the Corporation of Clyde. The plan was to carry the water across Waikerikeri Creek in a flume and feed it into Holt's race.

As far as is know the water was never used by Clyde Borough. Instead the water was leased to miners working at Muttontown Point and elsewhere along the riverbank. It is believed that the stretch of race from

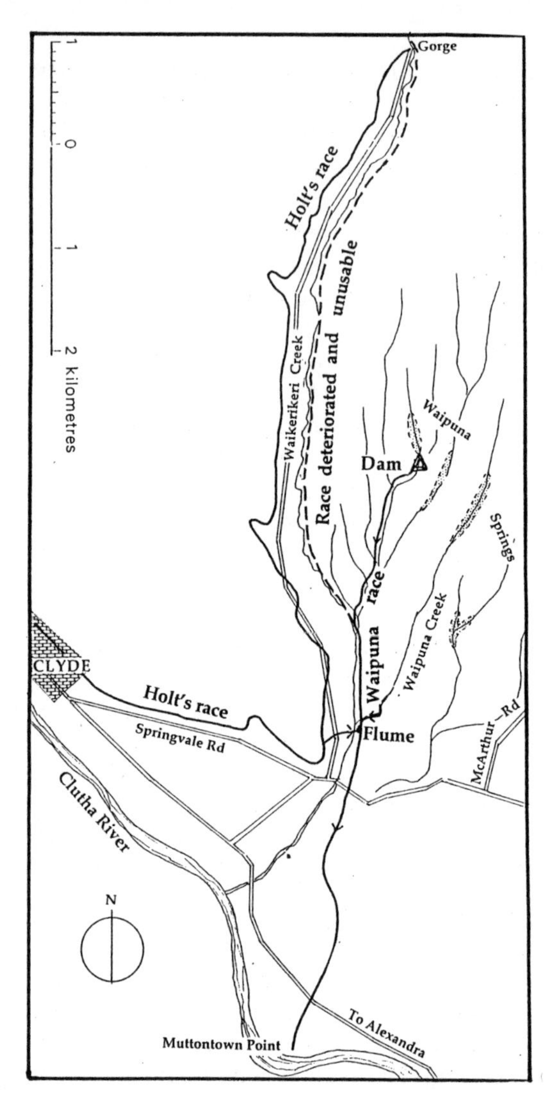

Figure 10. 7. The so-called Waipuna race owned by Clyde Borough as it was when taken over by Feraud in 1878.

the gorge down to the intake of the water from the Springs, was very difficult to maintain, and had in fact been allowed to fall into disrepair. Feraud said that as a recent Borough Councillor he had himself sponsored a resolution to have the race repaired, but it had not been done.

This gave the defence lawyer a chance to raise an interesting point, namely that in law 'A man can take no action against his own wrong,' so Feraud as a councillor could not bring a case against the council. This

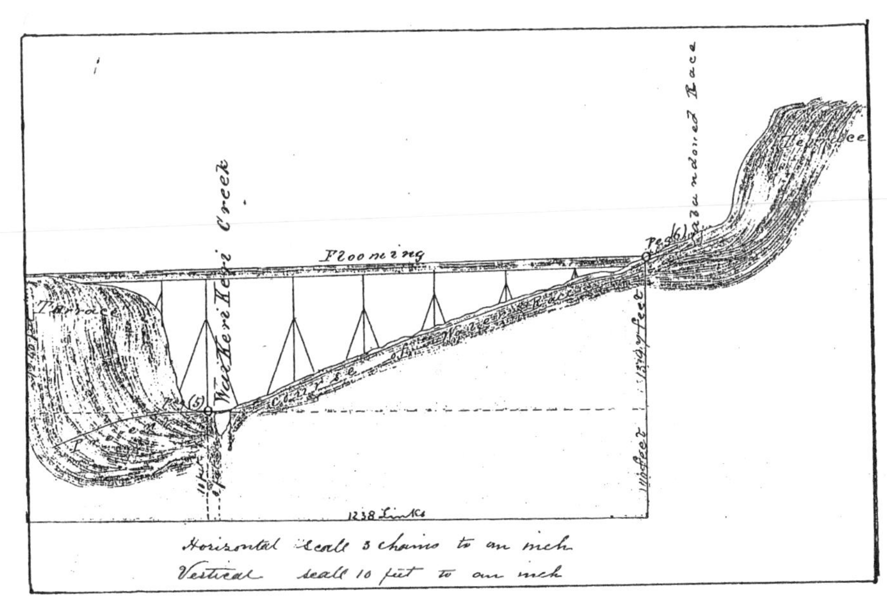

Figure 10. 8. This is the surveyor's sketch of the' floom' to be erected across Waikerikeri Creek to carry water from Waipuna Springs into Holt's race. The structure is 1,238 links (250 m) in length and 13.97 feet (4.2 m) high at the highest point. (Adapted from Mining Plan Otago S. O. 4998 courtesy LINZ)

resulted in an investigation of election documents and to everyone's surprise, and no doubt Feraud's relief, it was found that Feraud's nomination paper for the Borough Council had been sent in a day late and was therefore invalid. Since the election, Feraud had legally not been a councillor, therefore the legal point failed. Next day the council was advertising to fill its newly discovered vacancy.

Meanwhile the warden ruled that the race had been abandoned, and he therefore cancelled the licence. He could not transfer the water right directly to Feraud, but did not discourage Feraud from applying for it.[19]

Within a week of the judgement, Feraud had lodged his application for three heads of water from Waikerikeri Creek, and advertised the route of his race in the paper. The water was to be used for domestic, industrial, irrigation and finally, mining (when not required for the other purposes). The race would commence at the mouth of the gorge of Waikerikeri Creek and terminate on the banks of the Clutha River above the hospital.

The application for three heads from Waikerikeri Creek was heard in the Warden's Court over several days in February and March 1878. Objections from the Waikerikeri miners were thrown out on a technicality and the application was approved.[20]

Battle Over

The long battle for water was over. Well, not quite. When Feraud came to examine his newly acquired water supply, he found the Golden Gate party were illegally diverting the water from the Waikerikeri Creek into their race. He took them to Warden's Court but the warden, sympathetic to the struggles of this unfortunate group of hard-working men, dismissed the case. But they didn't pinch any more of Feraud's hard-won water.

After years of battling, the two antagonists James Holt and Jean Feraud, apparently got together in early 1878 to work out a deal about joint use of Holt's race. This suited Feraud as it saved him constructing a new race. For the use of his race, Feraud gave Holt the use of his water for the nine months of the year when he was not irrigating. During the summer slack period at the coalmine, Feraud was able to use most of the water in the race for his irrigation.

The question must be asked as to why this eminently sensible arrangement could not have been implemented in the first place, and so saved 10 years of hassle and expense.

There was a sequel to this convenient arrangement. In May 1891, James Bodkin, the then owner of 'Monte Christo' and who had taken over Feraud's water right, asked to exchange the title to the water race for a new title under a new Act. James Holt objected, pointing out that he had consented to J. D. Feraud using it in return for nine months supply of water. Now Bodkin refused to continue this arrangement, and therefore Holt, who pointed out that he still owned the race, had revoked his consent and would not in future allow any water other than his own to be carried in the race. The Warden granted a new title for the race and ruled that Holt could not now, after the elapsed time, refuse permission to use his race.[21]

Two months after this court case James Holt died as a result of an accident in his Dairy Creek coal mine.

Notes

1. Otago Provincial Government Gazette November 25, 1863 pp. 521-22.
2. *Otago Daily Times* 29 August 1864.
3. *Otago Daily Times* 8 October 1864.
4. *Otago Daily Times* 14 March 1865.
5. *Dunstan Times* 21 April 1871.
6. National Archives Dunedin Regional Office AAJQ/D98 4/8 127i.
7. *Otago Daily Times* 14 December 1872.
8. Votes and Proceedings Session XXX 1872 p 95. Feraud had been granted a lease over a 42 acre block as late as 1870.
9. *Otago Daily Times* 20 December 1872.
10. *Otago Daily Times* 21 December 1872.
11. The Otago Waste Lands Act 1866 Clause 124 '. . . the Board may refuse to

sell such land and in such case the Board shall if required by the lessee cancel the lease of such lands and pay the value of improvements. . .' The prohibition on sale applied to the Waikerikeri Valley and its adjacent auriferous lands. It did not prevent Feraud freeholding two sections in 'Monte Christo' (sections 43 and 59 totalling 25 acres) in 1874 and 1875.

12. *Otago Daily Times* 1 January 1873.

13. *Otago Daily Times* 6 February 1873.

14. *Tuapeka Times* 18 September 1873. Feraud had been declared bankrupt but was allowed to continue farming. He presented his case to the Inquiry on behalf of the Trustee, Mr R. H. Leary,

15. *Dunstan Times* 9 September 1873.

16. *Clyde Water Works Empowering Ordinance 1873.* Ordinances of the Province of Otago Session XXXII 1873.

17. *Dunstan Times* 5 June 1874.

18. Gilkinson 1936 p.196.

19. *Dunstan Times* 25 January 1878

20. *Dunstan Times* 8 February 1878.

21. *Dunstan Times* 12 June 1891.

11.

WATER ON SUFFERANCE

—Clyde's water Supply

The only water available to the first settlers in the tent township at the Dunstan was the river, and it was down a steep, 30 metre-high bank. If they didn't want to collect it themselves they had to pay someone to deliver it, by the barrel-full, to their door. So it was understandable that they would greet with satisfaction and relief the arrival of a water race in the township.

James Holt and his mate Roscoe, had originally planned a water race to deliver two heads of water from Waikerikeri Creek to miners at Muttontown Point on the banks of the Clutha River, but had changed their minds and diverted the race into Clyde (see Figure 10. 1). Its purpose now was to deliver water to work the pumps at the newly established coal mine on the banks of the river.

On its way to the mine the race passed along the back (the northern side) of Clyde township to its western end where it turned at right angles towards the river, and then reversed its original direction as it descended to the coal mine on the beach below the town. Here the water turned a small wheel working the pumps at Morgan's coal pit. The reason for this rather circuitous route through the town is unclear. Perhaps it was a generous gesture to allow as many people as possible access to the race, but perhaps it also ensured that a large number of citizens would come forward to defend their water supply in any future legal battles. It is interesting to speculate what arrangement Holt had with the townspeople for paying for this water. Holt does not appear to be the kind of man given to generosity, but at this stage there was no organised body in Clyde capable of acting as his agent.

The convenience of having a supply of water constantly flowing through the centre of town was no doubt much appreciated by the citizens of Clyde. In fact, according to the *Otago Daily Times*,[1] the supply was far in advance of that of any town in the province. Nevertheless, according to the Gold Fields Act current at the time, it was illegal. Water was for

Figure 11. 1. Between Clyde and the cemetery, traces of Holt's water race are still visible below the prominent 'Government' race.

mining only, with the understanding that miners could take a small amount for their own domestic use. It was not envisaged that the race water would be used to supply a township and it is very doubtful whether driving coal-mining machinery was a legitimate use of the water.

Be that as it may, Warden Keddell, almost certainly without legal right, had endorsed Holt's water race licence with the words 'Granted for domestic purposes for the town of the Dunstan,' and this apparently was taken to include the operation of the coal mine.

In April 1864 Holt took over the Dunstan Coal Mine, on which the town depended for its fuel, and he immediately improved its output. He installed efficient water wheel-driven pumps which had to be kept going continuously as the mine was located on the beach of the Clutha River, and the shaft went far below water level.

Feraud Threatens Water Supply

When orchardist and market gardener Jean Feraud threatened to construct another race to draw water from Waikerikeri Creek, the good folk of Clyde feared for the future of their water supply, which they fully realised was on shaky ground legally. A 'memorial' was sent from a public meeting to the Superintendent of the Province asking for a secure water supply. This had the effect of officially informing the Government of Clyde's illegal water supply and meant that the Superintendent, although probably sympathetic to Clyde's plight, could in his reply give only the

Figure 11. 2. This 1903 photograph shows Holt's race (extreme left), now used mainly for irrigation, still running along the back of Clyde township, and turning at right angles towards the main street where the water was used to flush the gutters.

legal position. He warned that when the Waikerikeri Valley, as part of a newly-established Agricultural Reserve, was settled, the farmers would be entitled to the use of the water in the creek, and the Clyde water would have to be restored to its former bed.[2] Neither the Superintendent nor the Provincial Government had the authority to override the Act to grant water to a township.

Clyde Borough

Clyde was declared a municipality in April 1866, and a couple of years later the Borough Council made a deal with Holt. It bought half of his water for £100, entitling the Borough to a constant flow of one head, but this did not mean that the municipality had acquired a legal water right, nor even a share in the ownership of Holt's race. The council also generously agreed to pay Holt £30 each year as its share towards the maintenance of the race. This arrangement persisted for some years even after the Borough was merged into Vincent County in 1878.

Beginning in 1870, Feraud stepped up his campaign through the Warden's Court to obtain the share of the water from Waikerikeri Creek to which, as a settler on an Agricultural Reserve, he firmly believed he was entitled. In 1870 the warden made an important Order. Under the Act a warden was able to direct, when awarding a water right, that a certain amount of water be allowed to continue to flow down a stream 'for general use.' Now he ordered that one head of water should at all times be supplied to Clyde for domestic use and to drive the coal mining machinery. This was as close as the warden could go towards granting the town a water right.

With each court case Feraud also made small gains in his fight to secure water, and the Borough Council became increasingly worried, in spite of Feraud's repeated assurances that if he were granted the water right, Clyde would still receive its water.

Largely as a result of requests from the Otago Provincial Government, which was under pressure by a number of towns having water troubles because of the restrictions of the Gold Fields Act, the Central Government acted. In October 1872 it passed the Municipal Corporations Water Supply Act which gave Borough Councils authority to buy or construct waterworks, and power to borrow money to pay for them. A separate Ordinance had to be passed by the Provincial Government for each Borough before the Act could be applied to it.

Clyde Buys Waipuna Race

In the face of the continuing aggressive attacks by Feraud on Holt's (and Clyde's) water, Clyde Borough Council decided to take out insurance. Anticipating the passing of the authorising Ordinance, it bought the Waipuna water race from Joseph Hastie.[3] Although now called the 'Waipuna race,' it had originally been constructed by Andrew Dalziel in 1864 to bring water from the gorge of Waikerikeri Creek to miners at Muttontown Point. The intake was above that of Holt's race, and the water right had second priority to that of Holt.[4] By 1873, the stretch from the gorge down to the junction of Waipuna Creek had not been used for some years and was in bad repair. However, it was not the race that attracted the Borough Council, but the water right from Waikerikeri Creek that still went with it.

Water from tributaries on the eastern side of the Waikerikeri Valley, fed by the Waipuna Springs, was still being led into the lower reaches of Dalziel's race and used by farmers, including Feraud.

Waipuna Springs

Although the flow from individual springs was small, the total from the dozens of these springs was in great demand by early miners because the supply was reliable. The Bladier brothers used water from the springs to irrigate their newly established market garden, and later it provided Feraud with his only source of water until he was successful in acquiring some from Waikerikeri Creek. Springs are not uncommon in the shallow gullies that dissect the gently sloping surface of the extensive high terrace which lies eastwards of Waikerikeri Valley. The reason for their presence is quite simple.

Figure 11. 3. Waipuna springs from the air. The springs emerge in small gullies which are tributaries of Waikerikeri Creek.

Although the high terrace appears to be formed from coarse gravelly and bouldery alluvium, this is only a surface layer a few metres thick. The alluvium is underlain by compact fine ·sands, silts and clays laid down perhaps 20 million years ago on the floor of a very large fresh water lake that extended over much of Central Otago. These fine sediments are only slowly permeable, so rainwater and water from small streams flowing out on to the terrace from the Dunstan Mountains quickly soaks through the porous alluvium until it is held up by the underlying fine sediments where it forms a 'perched watertable.' The water moves slowly down the gentle slope of the surface of the fine sediments until it finds outlets. The outlet may be alongs the sides of gullies formed during some wetter period of the past or the springs may themselves have formed the gullies by a process of 'spring sapping.' This is a process by which water bubbling out of a bank undermines the ground above, which collapses into the spring. The water washes away the fallen material, and as the process is repeated, a gully is slowly formed and extends towards the source of the water.

At Waipuna the springs form 'spring lines' along the sides of the gullies some distance above their floor. The spring line marks the junction of the alluvium and the underlying fine sediments that form the floor of the gullies.

Feraud objects to use of Waipuna water.

As Holt's race was still supplying the head of water purchased some years before, Clyde Borough did not immediately need the water from the newly purchased Waipuna race. In case it were required, however, an application was made to extend the race across the Waikerikeri Creek by means of a long fluming so that the water could be fed into Holt's race. To

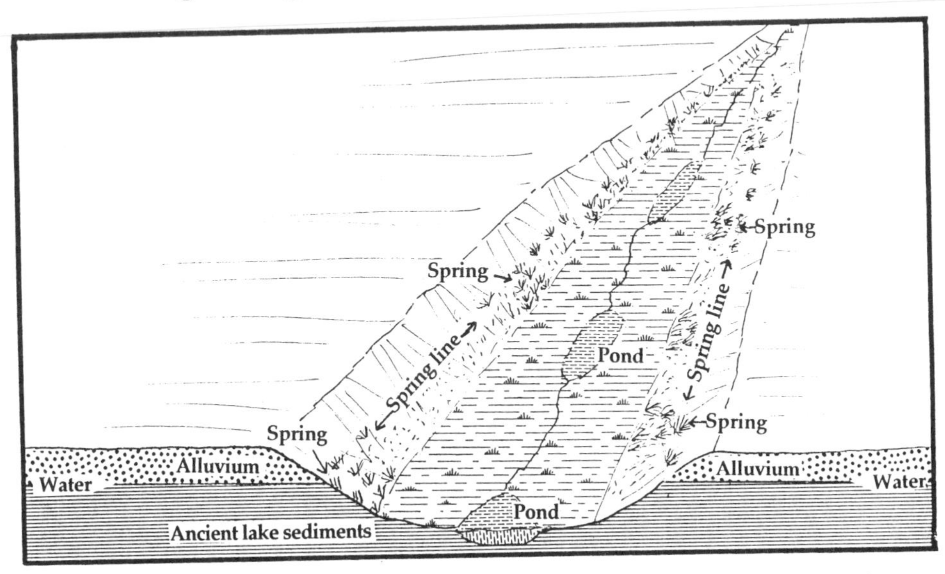

Figure 11. 4. A cross-section of the Waipuna springs shows how water, having soaked through the porous surface gravels, flows along the surface of .the underlying tight sediments to emerge in gullies as springs.

avoid any further embroilment in the meaning of 'domestic use,' Council gave the reason for the alteration as 'gold mining purposes' and indeed, the water was let out on tender to gold miners.

Feraud objected to the proposed alteration of the Waipuna race, pointing out that the Clyde Corporation had no charter for gold mining; that not being a mining company it could not make an application under the Gold Fields Act, and that the Borough Council had illegally obtained a certificate of transfer. Finally, as the race, contrary to the Act, had been unused by the council for a long time, Feraud asked that the transfer be cancelled and the council's rights be declared forfeited.

However, the lawyer defending the Corporation, pointed out that if the gold miners who had tendered for the water had not used it, it was not the Borough's fault. Furthermore, contrary to Feraud's view, the 1873 Clyde Water Works Empowering Ordinance did give the Borough Council the right to acquire water rights. Feraud lost the case.[5]

It turned out that 1874 was to be a very busy year for the Clyde Borough Council. With a rush of blood to its head, it decided to raise a

Figure 11. 5. This photograph, taken on Bruces Hill several kilometres south-east of Waipuna springs, shows the same formation of alluvium over old lake sediments. Water emerging at the interface between the gravel and sediments, is encouraging a line of vegetation.

loan of £2,000 'to construct water-works for supplying the Borough with water', and the necessary advertisements were placed in the newspapers.[6] Certainly the Council had the £200 to pay for Hastie's race, but what its plans were for spending the remainder of the money are no longer known. There are no further reports about the proposed 'water-works' and it is likely that the application for the loan was turned down by Central Government.

Feraud Wins

It was 1878 before Feraud returned to the attack. He recognised the Waikerikeri Creek-Waipuna race as the weak spot, and applied to the Warden's Court for cancellation of the licence, as the race from Waikerikeri Creek had not been used for mining for two years, and had lain abandoned and destroyed for more than a year. At long last he was successful, and the licence for one head from Waikerikeri Creek held by the Clyde Corporation was cancelled.

Within a fortnight Feraud had applied for the licence and it was granted.

Did Feraud Steal Clyde's Water?

Robert Gilkison, writing of Feraud in 1930, said:

> . . . his one bad act was when he broke the trust reposed in him as a public man and captured for himself a [water] right which belonged to the

> people who elected him. . . He chanced to be elected Mayor of Clyde, and in that position learned that the town water right was not legally held. Feraud resigned his position, and then took action in the Warden's Court, the result of which was that the Clyde right was cancelled and he himself obtained a licence for two heads of water from the Waikerikeri Creek. From this position of trust he should not have been allowed to benefit himself by such an action. . .[7]

This is a surprising statement from a lawyer, who had begun to practice his profession in Clyde only a few years after the departure of Feraud, and so must have had access to the court records and must have been familiar with the facts of the matter. In the preface to his book he claims that he has 'endeavoured to set down a truthful and accurate account.' It

Figure 11. 6. One of several dozen springs in the gullies neighbouring.Waipuna Creek.

is not that the individual facts mentioned are in themselves inaccurate, but the way they are presented is biased against Feraud.

For instance, at no stage did Feraud interfere with the town water supply. Certainly, Feraud had resigned in the middle of his second term as mayor, but it was over a row about staff discipline at the hospital in which his fellow councillors would not support his actions. It is carrying speculation too far to assume that he engineered this row so that he would have an excuse for resigning, in order to grab the town's water supply. Clyde's illegal use of water (according to the Gold Fields Act of the time) was well known to most people, including the Superintendent of the Province, so it was not something that Feraud suddenly discovered when he became mayor. Four years elapsed between Feraud's resignation and the beginning of his series of court battles to obtain water from Waikerikeri Creek, and he offered to continue to supply the town with water if and when he gained Holt's water right.

It was 11 years after his resignation from the mayoralty that he finally attacked and captured Clyde's water right. But this was not the town water supply—it was a race and water right acquired by the Clyde Corporation as an emergency supply years before, and his attack was not based on any illegality of the water right. It was based on the well-understood law of forfeiture — if a water race was not used for a month, it could be declared forfeited and the licence cancelled. The town had never used the water race and water right, and the stretch from Waikerikeri gorge to Waipuna Creek had not been maintained. Indeed, Feraud, during his last term as councillor, had sponsored a resolution to have the race put in order. The warden ruled the race and water right to be forfeited through lack of use.

Figure 11.7. L. D. Macgeorge was the Vincent County engineer who wrote a Damning report on the Clyde water supply.

So, in spite of Gilkison's[8] opinion, which has become part of the folk-lore of Central Otago, it seems that the Clyde Borough Council had no one to blame but itself for the loss of its race and the water right from Waikerikeri Creek. Perhaps the Council had lost interest in water—in a few months it was to be abolished and the administration of the township was merged with Vincent County. It is true that when Feraud finally won the right to water from Waikerikeri Creek after eight years of battling, he was awarded two heads of water from the creek.

Vincent County was formed in 1876, but it was not until April 1878 that the Borough of Clyde formally merged with the County, and a Town Trust was formed to administer certain assets such as the Library, Town Hall and the water right.

The county was busy with big projects such as the Alexandra Bridge, and it wasn't until these were completed that the engineer was able to turn his attention to the Clyde water supply.

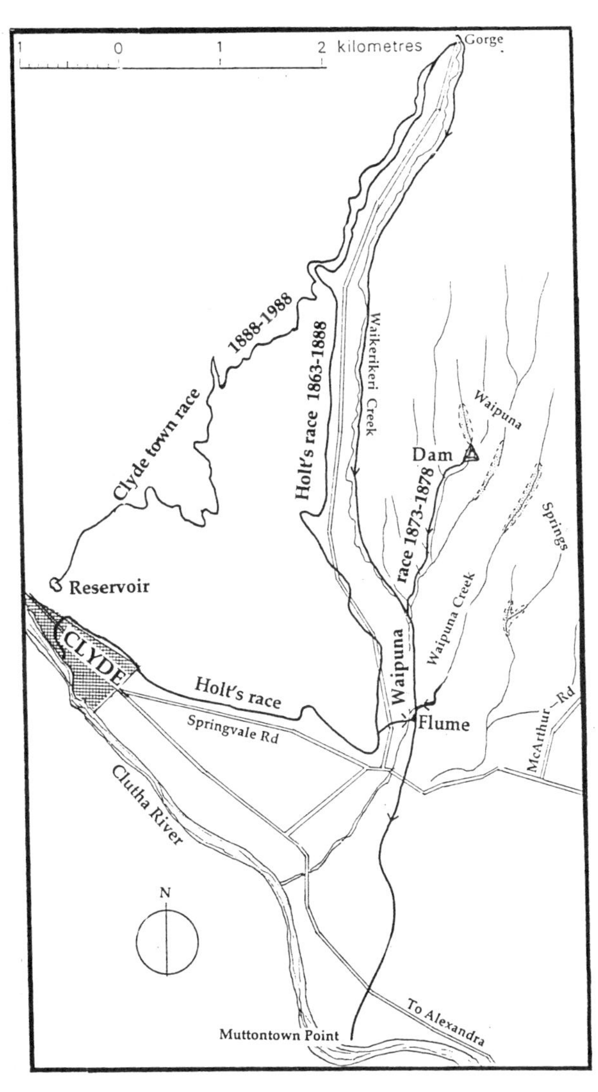

Figure 11. 8. A summary map showing the various water races that supplied Clyde with water over a period of 130 years.

Macgeorge's Report

The County Engineer, L. D. Macgeorge, was asked to inspect Holt's race in 1883 and report.[9] His was a shocking report. Had the water been delivered to Clyde in the same condition as it entered the race, there would have been no concern. It was, however, contaminated to such an extent on its journey to the township that it was quite unfit for domestic purposes and especially for drinking:

> Within 1/4 of a mile of the head there is a stable not more than 10 or 12 ft from the banks. It is on slightly sloping ground, and a drain is cut from the doorway into the race, leading urinal and other matter directly into it. At the time I inspected it there was also a large heap of manure piled up against the wall, its outer edge being no more than a foot or two from the water.

Further down the road the race ran through Kelliher's cowyard and was defiled in much the same manner.Even within the boundaries of the town the race

> . . . passes just beneath a slaughter yard. The yard is on the slope of the Terrace, and consequently during heavy rain a large quantity of filth and noxious manner must of necessity be washed into it.

Macgeorge went on to suggest that the Borough's water could be conveyed by a new race to the top of the terrace above the town, or possibly an arrangement made with the Golden Gate Company to use one of their races. Another option would be for Council to abandon the Waikerikeri water and buy water from the Golden Gate Company. The engineer suggested that a reservoir should be built in a convenient hollow just below the top of the terrace, and from there he would pipe the water down into the town which would be reticulated with water mains.

Macgeorge's estimated cost for the scheme was £1,675 that included £300 for constructing eight miles of race. But nothing was to happen for a few years.

Fifteen years before, in 1868, Clyde Borough Council had bought a head of water from James Holt,and at the same time agreed to pay him £30 a year as a share towards maintaining the race. Each year this money was paid over, and even after the Borough was merged with the County it was continued—until 1883. Holt's account for £60 (two year's fee) had arrived on the County Council's table. It was resolved 'that the account be not entertained.' When this news was conveyed to Holt he threatened to sue. Council decided to take legal advice.

Meanwhile Holt's bill had reached £120 and a committee of councillors was appointed to look at the whole matter. It reported in May 1886 that Holt had no legal claim for the money, and recommended a sum of £40 be paid without prejudice to liquidate the alleged debt from 1881, and thereafter £10 year be paid for keeping the race in proper repair, and maintaining a flow of one head of water in the part of the race that passed through the town.[10]

Numerous letters to the paper made it clear what ratepayers thought about the supply. One asked if the water supply really came from:

> that filthy looking race which flows from the back of the town and thence through the streets collecting the drainage and sewerage water in its course.

The County Solicitor complained of pigs wallowing in the race in 1886, and this caused the County Council to resolve that it was prepared to supply, to the extent of £900, a feasible water scheme for Clyde. Interest on the borrowed money was to be covered by a special rate on those who benefited.

Finally in January 1888, the engineer was instructed to make a detailed survey of a new water race for bringing water to Clyde, and to have tenders ready for calling at the next meeting. It was May, though, before tenders were called.

The New Race

The new race was to be constructed from the old intake at the mouth of the Waikerikeri Gorge, across the terraces and downlands, and terminating at the top of the high terrace overlooking Clyde. The detailed specifications[11] are of interest in that they tell us something of the technique of race building at the time.

The total length of the race was to be 8 miles (13.5 km) of which 590 feet (180 m) would be in wooden fluming, mainly of kauri. To guide construction, the engineer drove in pegs every 130 feet (40 m) along the centre line of the race. The race was to be excavated 15 inches (37.5 cms) below the ground line of each peg, and to be two feet (60 cms) wide at the bottom and 2 feet 6 inches (75 cms) at the surface. Fall to be one foot in 350 feet (1 m in 350 m).

Where the race crossed a low spot, or an outside wall required building up along a slope, the banks were to be built of sods rammed or rolled and twice soaked with water. Walls were to be at least 18 inches (45 cms) thick, and within the channel the outside of curves were to be protected with flat stones on edge. Two stone culverts had to be built to allow storm water to pass under the race. From the top of the terrace a race closely lined ('pitched') with rounded stones, was led down to a natural hollow about 200 feet (65 m) above the town, where it was proposed to build a reservoir capable of holding two months supply of water.

The contractor had to 'bring the water with him' as he built the race, and was to finish the job in four months. He had to maintain the race for three months, repairing all leaks, and finally, he was not to employ any Chinamen. The tender of Herbert Harding of Clyde for £269 17s 11d was accepted.

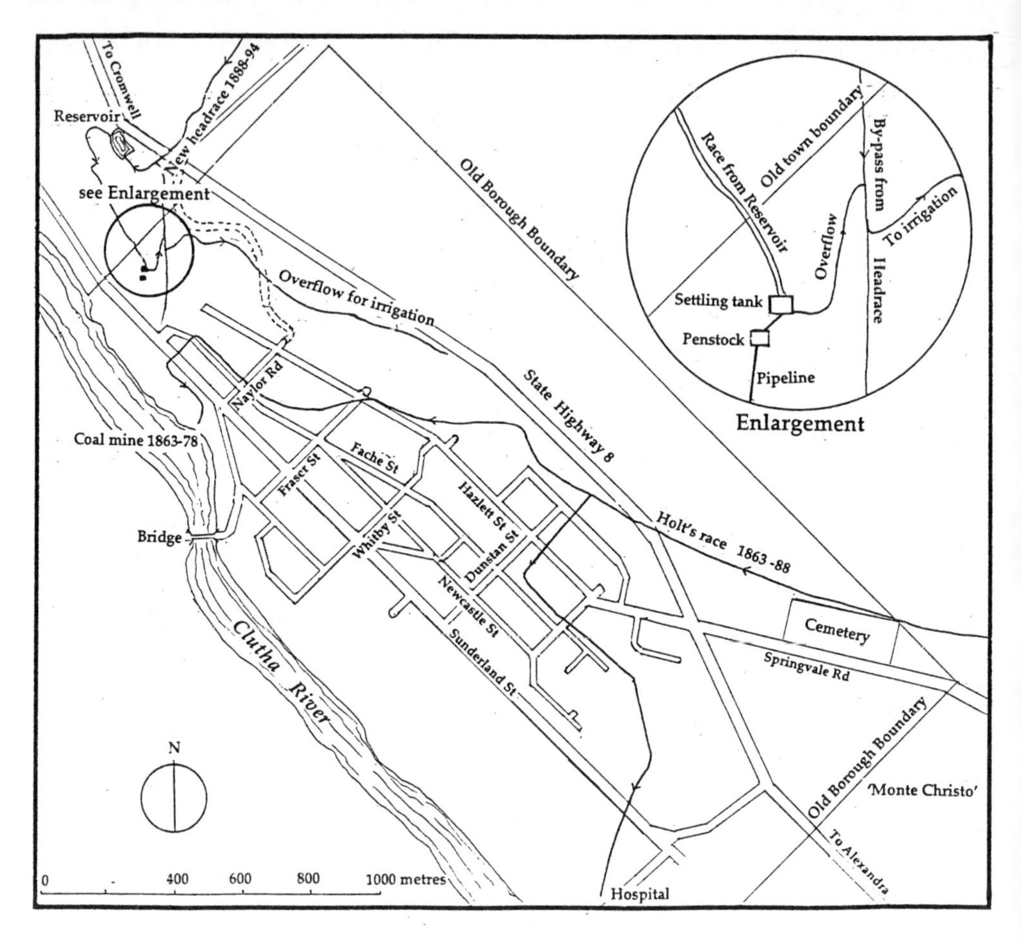

Figure 11.9. Plan of Clyde showing route of Holt's race through the township and details of the water works installed in 1903.

The reservoir was not built at this time and water from the race was fed directly into a pipeline. About two kilometres of four-inch (100 mm) diameter pipes distributed water through the streets where a number of upstands with taps were installed.

Apparently pure water still eluded the townspeople, and there was still criticism. A newspaper correspondent talked[12] of drinking mud, and he put forward a financial plan for a new scheme which involved building a reservoir and increased reticulation. A councillor described the existing scheme as 'unsatisfactory, inadequate and unsanitary.[13]

The problem was money. The Town Trust still owned the water right and operated the water supply scheme, and was responsible for paying off the various loans raised by the County on the Trust's behalf. In January 1892 the Clyde Water District was formed and a rate struck to help pay off previous loans and finance future expenditure. The executors of James Holt's estate offered his race and two heads of first priority water for £600 and this offer was taken up. Apparently it was felt necessary to have this water for irrigation within the township. A loan of £800 was raised for this and ancillary works.[14]

The Town Trust was soon in arrears with repayments of these loans and the County was forced to come to the rescue. It raised a £12,000 loan to take care of the outstanding debts, and to provide for a proper reservoir and reticulation scheme.[15]

The Government, through the Public Works Department, contributed £1,121 11s 2d towards the scheme.

New Waterworks

It was a proud day for the citizens of Clyde, we are told, when their new water supply scheme was at last opened on 24 September 1903, just two months before the Alexandra scheme was also opened. The Clyde scheme was designed by the County Engineer, G. L. Cuthbertson. Water from the Waikerikeri Creek race was first impounded in a reservoir situated in the natural hollow just below the crest of the terrace. An open race took the water 100 yards to a settling tank. A short distance away was the penstock, excavated from solid rock and cement lined, which led the water into the pipeline that fed the towns reticulation of water mains.[16]

Figure 11. 10. The Clyde town race, now abandoned and overgrown, supplied water from Waikerikeri gorge for almost exactly 100 years.

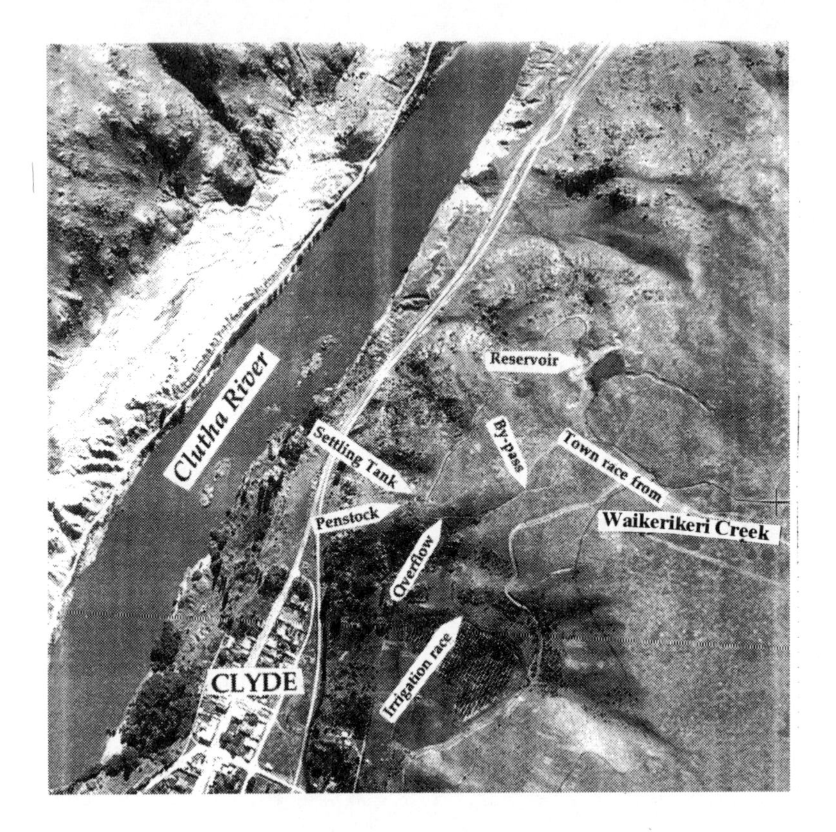

Figure 11.11. An air photograph taken in 1949, shows details of the water works such as the race from Waikerikeri Creek, reservoir, settling tank and penstock at the head of the pipeline. Water by-passed from the race, and overflow from the tank, was used for irrigation.

This scheme, with necessary adjustments and renewals, served the town for 80 years and was only replaced in 1981 when the preliminary work for the Clyde dam began. Because a new highway was to skirt the huge dam and to cut through the old reservoir, Clyde was supplied from a waterworks built to supply the needs of the dam construction. The town is now served with water supplied directly from the dam.

NOTES

1. *Otago Daily Times* 10 July 1863.
2. *Otago Daily Times* 8 October 1864 .
3. *Tuapeka Times* 8 May 1873 and *Otago Daily Times* 30 April 1873.
4. Granted 17 November 1865. Warden's Court, Clyde. National Archives, Dunedin Regional Office.
5. *Dunstan Times* 3 April 1874.
6. *Dunstan Times* 5 June 1874. Cannot be checked as the relevant issues of the *Dunstan Times* are missing.

7. *Early days in Central Otago*. 1930. p. 196.

8. Veitch 1976 pp. 55-60 sets out a much more balanced, but incomplete, account of the water affair.

9. *Dunstan Times* 26 October 1883.

10. Vincent County Council Minutes November 1883 to May 1886 Hocken Library.

11. Vincent County Council *Tender Book* Hocken Library.

12. *Dunstan Times* 14 December 1900.

13. J. H. Angus p. 104.

14. Vincent County Council Minutes March and July 1892.

15. J. H. Angus p.105.

16. *Otago Witness* 30 September 1903.

12.

EVER HOPEFUL

Waikerikeri Gold Diggings

When the Clutha River rose in mid-September 1862 and washed the miners from the Dunstan (Cromwell) Gorge, they carried their hunt for gold out into the surrounding countryside. One of the first places they prospected was the valley of the Waikerikeri Creek, which entered the Clutha River from the north by way of a steep-sided gully a few kilometres downstream from the Dunstan township. It was near the mouth of this gully that miners congregated to buy meat from the sheep stations which were slaughtering sheep to supply the men. So the collection of miners' tents became 'Muttontown,' the nearby gully became 'Muttontown Gully' and not far below the gully was 'Muttontown Point' where the Clutha River turned sharply to the east.

After leaving its mountain gorge, Waikerikeri Creek continues almost due south for eight kilometres towards the Clutha River in a flat-floored valley, half a kilometre wide and about 20 metres deep, cut into the high terrace of the Manuherikia Valley. It was disappointing to the prospectors to find that gold was sparse in the alluvium on the floor of this valley. Certainly, they found 'colours' of gold in the deep gravels, but the quantities that could be recovered with the primitive appliances available were simply not payable.[1]

Miners had better luck, however, in the valleys, or 'gullies' as they called them, of eight or nine small tributary streams that entered the main Waikerikeri valley from the slopes of the Dunstan Mountains. In time past, these tributary streams had cut deep, more or less parallel, flat-floored valleys across the high terrace before joining the main stream. Long, nearly flat-topped remnants of the terrace were left between the gullies.

Identifying the Gullies

These gullies were all given names that were used in applications to the Warden's Court, and are quoted in newspaper reports. These miners'

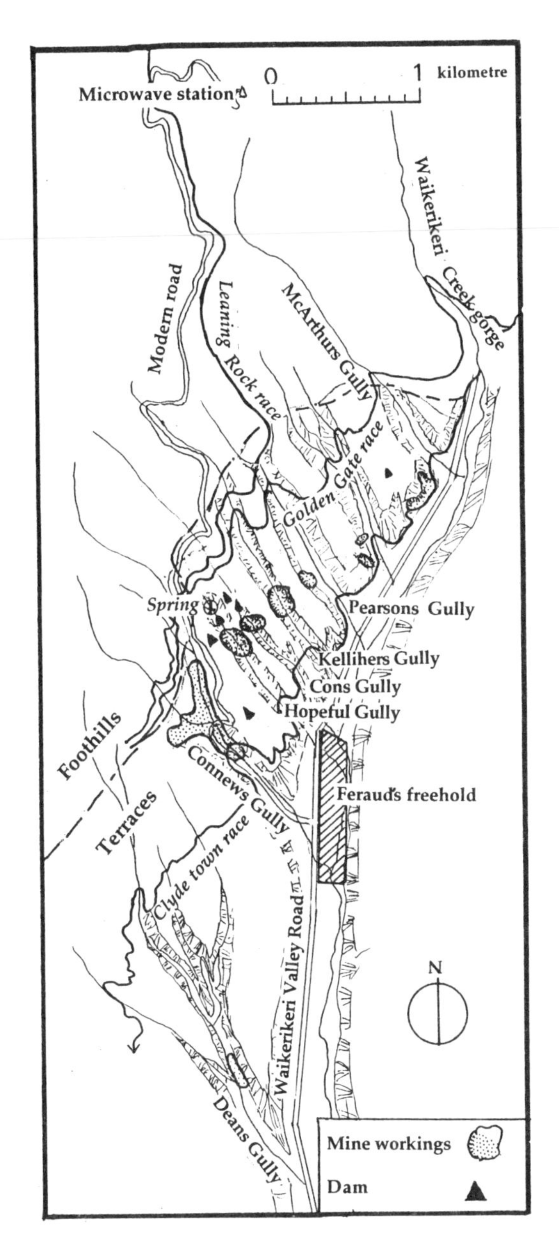

Figure 12. 1. The tributary gullies of Waikerikeri Creek showing gold workings, water races and dams. Also shown is the freehold property of J. D. Feraud which blocked the discharge of tailings and threatened the development of mining.

names do not appear on modern maps so it is difficult to locate the old claims, water races and other features referred to in these applications. To add to the difficulties, many of the names used were ephemeral, and were quickly abandoned or changed. With the help of Mr Sydney Athfield of Clyde, who remembered names used by his grandfather, who had farmed in the Waikerikeri Valley during the early 20th Century, it was possible to match most of the old names to the gullies.

Figure 12. 2. Air view of the terrace with gullies cut by small streams flowing from the foothills of the Dunstan Mountains to the Waikerikeri Creek (foreground). Gullies are Hopeful, (extreme left with trees), Cons, Kellihers, Pearsons (centre). Old name for the large gully to the right is not yet known.

One feature that can be located with some certainty is Connews Gully, because it is known where James Connew farmed. With the position of this gully established, it was possible to identify the gully adjacent to Connews as 'Hopeful Gully,' because there was an application to the Warden's Court for a water race from Hopeful Gully to Connews Gully that was only 400 yards (365 m) long. In another report 'Jays Gully' and 'Hanlons Gully' were described as adjacent, and an application by Con Kelliher for a claim in 'Jays Gully' confirmed that this was the old name for what was later called 'Cons Gully'. Similarly 'Hanlons Gully' later became 'Kellihers Gully.' It is assumed that names such as McLeods,

McBrides, Watsons, and Gilchrists Gullies (all names used in applications for claims) were ephemeral names used for a short time while the named person was working in a particular gully.

Mining the Gullies

By 1868, the floors of most of the gullies had been tested and almost all were found to have gold under the alluvium, especially in their upper reaches. In most, however, water for mining was inadequate. The normal flows of the streams entering these gullies from the hills was very small, and most of the water soaked into the gravels of the gully floors. Some reappeared lower down the gullies as springs that dried up during the summer heat and the winter frost. But by digging holes or making small dams, the miners were able to accumulate sufficient water from these springs and seepages to allow some mining by panning and cradling.

One gully however, attracted most mining activity because it had a permanent spring, which, small as it was, provided sufficient water for panning and cradling. Here the gold, in 'wash-dirt' (or simply 'wash') of sand and fine gravel, was lying on smooth 'clay'[2] under only shallow alluvium. The gully was wistfully named 'Hopeful Gully' because the miners were always hopeful someone with money or initiative would build a water race to supply them with adequate water. All they asked for was one sluice-head of water so they could carry out ground sluicing.

Figure 12. 3. The spring in Hopeful Gully which provided sufficient water for miners to pan and cradle the alluvium on the floor of the valley.

Mining activity in these gullies can be separated into several phases. First, there was the mining of the shallow alluvium on the gully floors. Because of the lack of water, activity was confined largely to Hopeful Gully with its spring.

A second phase began when two large water races were brought in from distant streams, and used by their owners for wholesale sluicing in Connews Gully. There was a third phase when one of the water race companies made water available for sale. This allowed a number of small groups of miners to begin sluicing into the terraces flanking the other gullies. Finally, when the depth of overburden became too great, drives were excavated along base of the base terrace alluvium to recover particularly rich wash.

Figure 12. 4. A remnant of the Leaning Rock race preserved below the modem road to a microwave station

The Big Races

In the early 1870s, miners were excited to learn that two separate groups were planning to bring water to the gullies by way of large water races. There was excitement too amongst the townspeople of Clyde who looked forward to a greatly increased number of miners on the diggings, with subsequent increase in business. Little wonder there was dismay and anger when it was made known that J. D. Feraud had freeholded land on the floor of Waikerikeri Valley. This meant it would be very difficult, if not impossible, to dispose of the large amounts of tailings from the extensive sluicing operations that would take place once the race water arrived. The resulting confrontation has been described in Chapter 10.

Leaning Rock Water Race

John Lindsay and James Robertson were granted a licence in March 1871 to cut a water race from Leaning Rock Creek, the largest stream falling into the Dunstan Gorge. It took about 18 months to complete the eight mile (13 km)-long race, which had to begin at an elevation of 1,700 metres to avoid the heads of the numerous large gullies etching the slopes of the gorge. The water was eventually discharged into the head of Pearsons Gully, a tributary of Waikerikeri Creek. Some distance down this gully the water was picked up by another race and conveyed to the 'Diggers Gully' (Hopeful Gully).

With the arrival of Lindsay's four sluice-heads, miners at last had sufficient water for ground sluicing the floors of the gullies. In this method, the 'wash' was thrown into a stream of water that had been directed through a ditch lined with flat stones. The gold was caught under the stones while the waste sand and gravel passed out of the ditch as 'tailings.' Every so often a 'wash-up' took place when the stones were lifted and the gold, and its accompanying black sand, was washed or swept out of the ditch and the gold panned or cradled off.

Although Lindsay and party made some water available to the miners in Hopeful Gully for a short time, it was their intention to mine themselves, so they carried the race across the terrace to the next gully lying to the south-west. It was in this much larger Connews Gully, as the miners called it, that Lindsay's party set about some serious mining. With the withdrawal of Lindsay's water from Hopeful Gully, mining there came to a standstill and the place became virtually abandoned.

Connews Gully

James Connew had been successful, in early 1867, in gaining an Agricultural Licence for 50 acres (20 h) over one of the largest of the tributary gullies, and he was now farming its flat floor. But gold had been found in this gully also, and by 1870 Thomas Jay and his partner had cut a water race to collect the water of the tiny stream before it soaked away into the gravel of the gully floor. With this water they began cradling on Connew's farm. They were soon joined by Robert Watson who cut a race to bring water over from the spring in Hopeful Gully.

Presumably it was the success of these miners that encouraged Lindsay and his partners to bring their Leaning Rock race into Connews Gully. There were constant changes in the partnership that owned the race. Newcomers who played an important role in the partnership were the brothers David and Walter Anderson, who took over James Robertson's half share, and John Lawson, who became the principal shareholder in

1874. At one stage John Lindsay sold out completely, but four years later bought back in again.[5]

Lindsay's party soon built a dam in Connews Gully and pegged out a four-acre claim. They were lucky: their first wash-up returned 100 oz of

Figure 12.5. Air view of Conroy's Gully (left) and Hopeful Gully (right).

gold. In a short time they had paid for the cost of the water race. Lindsay and his partners used their water carefully. From experience they knew that they could expect their full supply of water for only a limited period while the snow on the mountains was melting in the spring, so they used this full supply to strip off the overburden of barren gravel. Then when the water supply began to lessen in the summer, they used it to wash the exposed 'wash-dirt' through the gold-saving sluices.

In spite of the initial good returns, the gold turned out to be patchy. It was only large scale sluicing, such as carried out by Lindsay's party, which averaged out the returns and made mining payable.

Lindsay's party made an important discovery. Up on the high terrace between the gullies, they sank a shaft and found that after passing through 50 feet (15 m) of auriferous, but non-payable, coarse schist alluvium, they reached a layer of gold-bearing wash five feet (1.5 m) thick resting on grey 'clay.' Because of this discovery, mining activity in later years turned away from the alluvium on the floor of the gullies and concentrated on the high terrace remnants.

Golden Gate Water Race

Meanwhile another party of experienced miners and race builders was also thinking about an ambitious water race scheme. The McNally brothers, Pat and John had been mining at Blacks (now Ophir) and had brought in a water race from the Dunstan Mountains. They soon found that selling their surplus water to other miners was as profitable as mining. Now they were planning to cut a race along the foothills of the Dunstan Mountains to convey water from Chatto Creek to the eastern bank of the Clutha River, where there was a great deal of mining but a desperate shortage of water.

Calling themselves the Golden Gate Company, they began work on the race in mid-1872, but the job was much bigger than they had expected.[3] Most of the construction was through easy ground, but the main obstacle was the gorge of the Waikerikeri Creek. The race had to be constructed for about three-quarters of a mile (1 km) along the side of the precipitous gorge until it reached the level of the rising creek bed. After crossing the creek the race had then to be built down the opposite side which was, if anything, even more precipitous. For almost the total distance within the gorge the race had to be supported on stone walls, in places up to 20 feet (6 m) high.[4]

It was not until late 1875 that the race reached Connews Gully, and by this time the race builders, exhausted physically and financially, had given up any ideas of extending the race to the banks of the Clutha River. They were content to terminate their race where they were, and begin mining the several claims they had taken out.

When the Golden Gate Company began sluicing in late 1875, the only other party working on the field was that of Lindsay. All the others had drifted away, mainly owing to the lack of water. Although the original plan had been to sell its surplus water, the Golden Gate party was not in a position to do so, because its race was not yet connected to its intended source in the headwaters of Chatto Creek. The company was concentrating on working its own claims in Connews Gully with the limited amount of water collected from small mountain creeks picked up along the race.

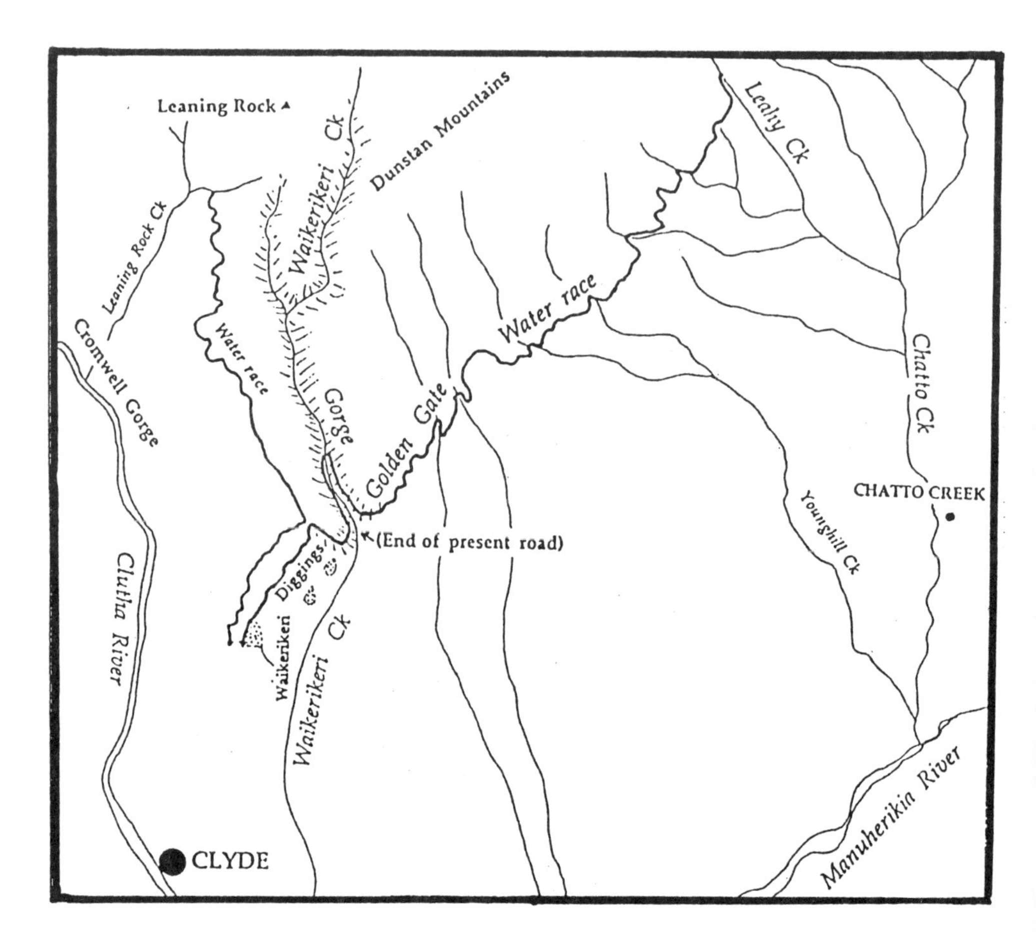

Figure 12.6. Map of Golden Gate and Leaning Rock water races.

The Golden Gate Company struggled on for a few years, but disillusioned by poor returns, beaten down by the physical work and the burden of the £8,000 cost of the race, the party began to break up. As the members left one by one, their shares were bought by B. R. Baird of Cromwell, who, in February 1878, formed a public company. With money now available, the first task was to complete the race back to Chatto Creek, and once this was achieved a large volume of water was available for sale at Waikerikeri Diggings. By October 1880 no fewer than eight parties had been enticed back and were at work using Golden Gate water.

Sluicing the Terraces

With sufficient water available, hydraulic sluicing became the favoured

Figure 12. 7. This photograph of the workings in Connews Gully was taken in 1905 by Professor James Park during the course of his geological survey of the district.

method of mining. Remembering the wash found at the bottom of the 50 foot shaft sunk on the terrace by Lindsay back in 1874, several parties began to sluice into the scarps of the terraces flanking the gullies. Because water had to be purchased from the Golden Gate company and was expensive, it had to be used economically. Small dams, built up on the terrace and filled overnight with water from the Golden Gate race, held sufficient water to allow several hours of sluicing next day. Pipes led down from the dam to sluice nozzles that played on the steep side of the terrace. Thirty or forty feet of coarse schist alluvium was quickly removed, exposing the wash of fine gravel and sand lying either on the blue-grey, old lake sediments or on a layer of small, brown sandstone (greywacke) pebbles. This layer of greywacke pebbles, weathered to the point of crumbling in the hands, was well known to the miners as forming the 'Maori Bottom.' By this they meant that the sandstone layer provided a resting place, or floor, (the 'bottom') for gold accumulation. The name 'Maori' was perhaps a reference to the colour.

As the sluicing advanced into the terraces, the overburden became thicker and the sluice-face higher so it became more economical to remove the gold-bearing wash by tunnelling. Some of the tunnels were of considerable length and had a number of side galleries. Wash was taken out of the mine in hand-pushed trolleys and washed through ground sluices or sluice boxes.

Well-known names of miners who worked on the terraces include Walter Anderson and O'Connell working in Hopeful Gully and later in Connews Gully, and Williams and Farrell and the Kelliher brothers in Jays Gully (later called Cons Gully). Clayton, and Kitto and Aitken worked in Hanlons Gully (Kellihers Gully)

Neil Nicholson was one of a well-known Central Otago mining family associated mainly with the many attempts to extract gold from the Fraser Basin behind the Old Man Range. Now, in September 1894, Nicholson, in partnership with John Leamy, began sluicing at Scrubby Gully at Springvale, using water brought by a five mile (8 km) branch race from the Golden Gate. It appears that at this time Nicholson and Leamy had taken over all of the Golden Gate water. It was not long before Nicholson had moved to Waipuna valley, a tributary valley joining the Waikerikeri Creek from the east just above the Dunedin road. Again he used water from the Golden Gate race.

Figure 12.8. The workings at Nicholson's claim overlook the road in the upper valley.

In September 1895 Nicholson moved his operations to the Waikerikeri Valley, where he had pegged out a claim overlooking the road at the lower end of the terrace flanking McArthur Gully (the gully adjacent to and parallel to the Waikerikeri Gorge). But a short time later he was appointed manager of the large Scandinavian Mine at St Bathans. The sluicing claim in Waikerikeri Valley was carried on for a time by a manager but in 1897 was closed down. Although Nicholson had control of the Golden Gate water, other miners working on the terraces spoke of his generosity in allowing them the use of the water on Sundays. This allowed them to sluice the wash they had excavated from their tunnels during the week.

With the liquidation of the Golden Gate Company in 1899, mining virtually came to a standstill. A number of the miners such as Anderson and Kelliher concentrated on the farms they had already established, whereas others, including Kitto and Aitken, took an interest in coal mining which was a growth industry serving the increasing number of gold dredges.

Today

The only signs of mining visible from the Waikerikeri Valley road are the extensive sluicings of Nicholson's claim on the end of a spur near the end of the road. The road rises over the fan of tailings that have been washed from these workings. The most extensive workings, the results of the activities of Lindsay and party and the Golden Gate Company, lie largely overgrown in the upper part of Connews Gully. Perhaps the most spectacular workings are the deep, vertically-sided gulches that have been sluiced opposite each other on both sides of Hopeful, Cons and Kellihers Gullies.[6] In Hopeful Gully some of the drives from which the gold-bearing wash was excavated, are still visible.

Figure 12. 9. Gold workings still visible in Hopeful Gully are the results of hydraulic sluicing and driving.

No one made a fortune at the Waikerikeri Diggings. The patchiness of the gold was notorious and gave the diggings a poor reputation. Miners would work for months for little gold and then just as they were going to give up and move on, they would strike a rich patch. They were encouraged to continue working with renewed enthusiasm, but for little return, until the next patch was struck. It was this reputation for patchiness that counted against the Waikerikeri Diggings becoming one of the more highly regarded diggings of Central Otago.

NOTES.

1. Forty years later the valley floor was completely pegged out for dredging claims, but fortunately, before the dredges were built, the ground was tested by intensive boring and sampling. This confirmed that the gravels of the valley floor were not worth dredging.
2. Although the bluish-grey 'bottom' on which the 'wash' lay was almost always described by miners as 'clay' the material is seldom a true clay. The sediment, which underlies the main Central Otago valleys such as the Upper Clutha, Manuherikia, Ida and Maniototo, consists mainly of fine sands with layers of clay which, near the base, often include coal seams. The presence of fresh-water fossils points to its origin as lake beds. It is often underlain by pure quartz gravel—the 'granite wash' of the miners.
3. For a much fuller description of the construction of the Golden Gate water race see J. D. McCraw, *Mountain Water and River Gold* Chapter 10.
4. These extensive stone walls may be seen after a short walk along a track leading up the gorge from the end of the Waikerikeri Valley road.
5. The two original partners in the water race, Lindsay and Robertson, owned two quarter-shares each, so each was able to sell one quarter -share and still retain a quarter-share. Over the 20 year life of the race no fewer than 12 different men held shares in the syndicate.
6. These old workings are on private property.

13.

FAILURES ON THE FRASER

Early prospectors expected the Fraser River to yield large quantities of gold, but they were disappointed. Although it was subjected to the whole range of gold mining techniques—panning, cradling, sluicing, hydraulic elevating, quartz crushing and dredging, the valley did not yield anything like the high returns of Butchers and Conroys Gullies on the eastern side of the Old Man Range. It is true, however, that the river and its tributaries provided a livelihood for a substantial number of individual miners, but none of the larger groups met with much success.

The Fraser River rises amongst the peat bogs and semi-permanent snow banks of a broad, high-altitude valley lying between the Old Man Range to the east and the ridge to the west, which is a southern extension of the Old Woman Range. The floor of this Fraser Basin, as it is called, rises to an altitude of over 1,400 metres and the Fraser River (or Earnscleugh River on early maps) descends from the Basin by way of a steep narrow gorge some 20 kilometres in length. Twice in the length of the gorge, the river's tumultuous descent is checked, as the gorge widens briefly into small basins partly filled with alluvium. One of these basins is now the site of the reservoir of the Fraser Dam, and the other just above the junction of the Fraser River with the Hawks Burn. When the river, now a substantial stream, finally emerges from the gorge it flows sedately across Earnscleugh Flat and joins the Clutha River a few kilometres above Alexandra.

Gold Discovery

On 11 November 1862, only 10 weeks after the Dunstan Rush, prospectors reported the discovery of gold in the big valley on the other side of the Old Man Range. Two parties, those of Davis and of Vallance (or Vallender) more or less simultaneously claimed double-sized claims as rewards for the discovery, and as these initial strikes were said to be 12 miles (19 km) apart, both were recognised as new finds.[1] It is probable that one of these discoveries was in the northern part of the Fraser Basin, and the other in the basin now partly occupied by the Fraser Dam although the straight-line distance between these point is only 11 kilometres (7 miles). The gold was heavy and coarse, but was patchy.

Two hundred men were soon at work in the upper reaches of the river and its tributaries, but the attractions of other gold discoveries at Cardrona and the Arrow River soon drew most of the miners away. The mining population was boosted after 1873 by numbers of Chinese, many of whom spent most of their mining lives in the Fraser Basin.

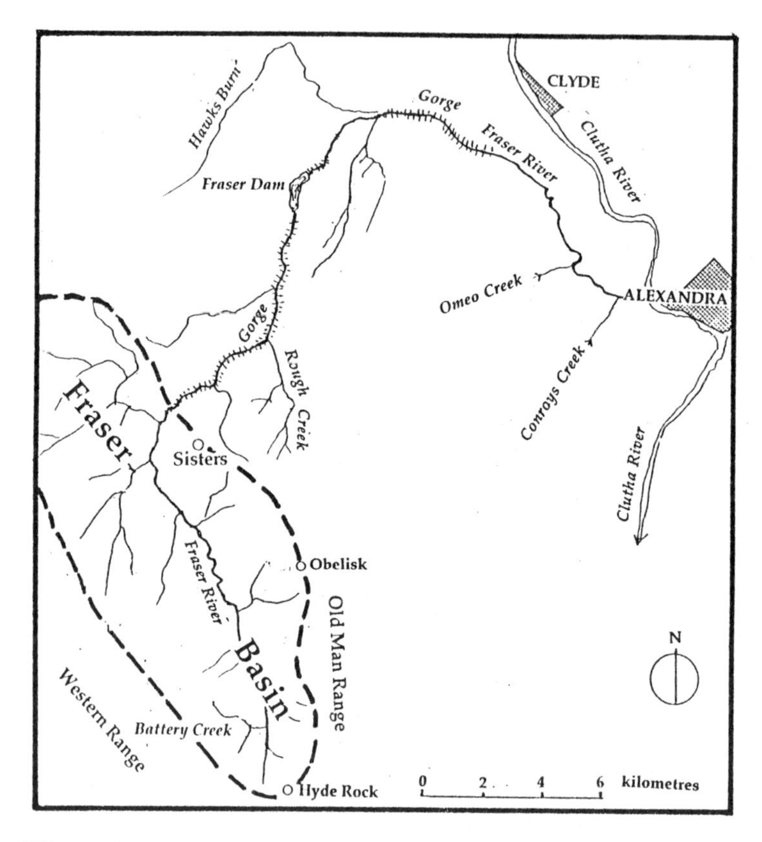

Figure 13. 1. Map of the Fraser River and the Fraser Basin.

THE FRASER BASIN

The Fraser Basin is about 12 kilometres long by about seven kilometres at its widest. At the head of the Basin a number of small streams, falling from the boggy slopes of the Old Man Range, and from the cliffs of glacial cirques in the western range, come together to form the Fraser River. The river meanders through a swampy flat with extensive peat bogs for about three kilometres before it enters a narrow shallow valley, which it follows for six kilometres before it broadens out into the lower or northern part of the Basin. Several large tributaries, especially from the western side, join the river here before it begins its descent from the Basin.

The Fraser Basin was, and still is, an inhospitable place. Its elevation, higher than Potters and Campbell Creek, meant that deep snow lay far into the spring, and the ground was frozen for long periods. Biting winds and blizzards at any time of the year made life uncomfortable and dangerous for the miners. Luckily, most had pulled out before the worst of the disastrous winter of 1863. It then became standard practice to abandon the Basin before the winter and not return until October or even later. Many miners used the Basin only for summer mining, when the Clutha River was running high and covering the beaches where they worked during the winter.

Among these summer miners was George Ratcliffe and his mates. George has left a description of how he was caught in a severe snowstorm on the summit of the Old Man Range during his journey into the Basin in November 1862.[2] He was fortunate to reach his objective safely, although his experience was but a taste of the hazards miners faced in this bleak locality.

Warden Robinson tells[3] of a charge of assault brought against miners as a result of a dispute over claim boundaries in the Fraser Basin. It was held by one of the parties that the marks deliniating the boundaries of a claim were no longer visible, so the claim could be taken over. Little wonder the pegs were not visible — they were covered by five feet of snow! The intending claim-jumpers held that the tussocks placed on top of the snow by the owners to mark the corners of their claim were not legal. In the same report the warden mentions another instance in which miners in the same locality were carrying on operations by tunnelling under several feet of snow.

Alpine Quartz Reef Company

Mining in the head, or southern part, of the Fraser Basin is associated with the Nicholson family. Charles Nicholson was from the Isle of Skye, and came via India and Australia to New Zealand in 1862. He established Nicholson's Hotel at Ettrick and ran it for 24 years, dabbling in mining investment now and then.[4] But when he retired, he devoted much of his time to mining and seemed to have an affinity for the Fraser Basin. One of his first ventures, however, was a spectacular failure.

With John Fry Kitchin, the recently retired manager and lessee of the vast Moa Flat Station, Nicholson formed a company with half a dozen friends and relatives to open up a supposedly gold-bearing reef that had been discovered right at the head of the Fraser Basin. A 20-acre claim was granted in January 1882, and no time was lost in carting in a 10-stamper battery, which cost £2,000, and erecting a 22 ft 5 in (6.8 m) diameter water wheel to drive it. Then there was silence. There were no reports of stone being crushed, let alone gold recovered. The Inspector of Mines, perhaps by way of explanation, pointed out that the high

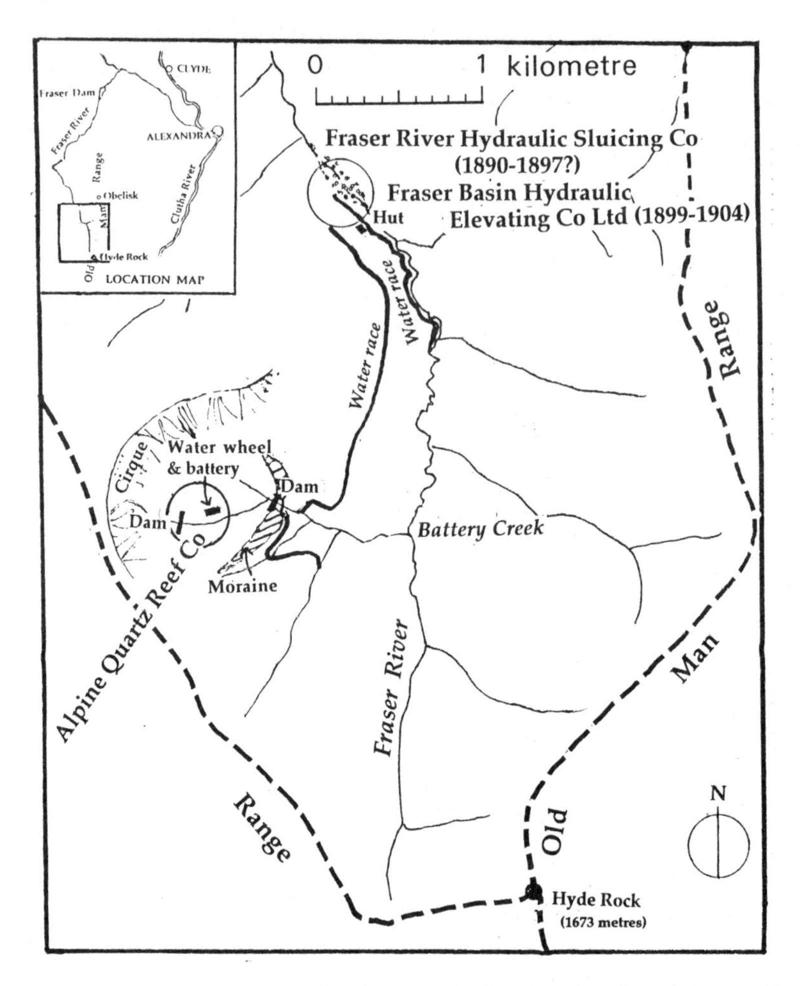

Figure 13. 2. Map of the head of the Fraser Basin showing the claims of the Alpine Quartz Reef Company, Fraser River Hydraulic Sluicing Company, and the later Fraser Basin Hydraulic Elevating Co Ltd.

elevation of the mine prevented operations being carried on for several months of the year. Years later, comments in newspapers filled out the story:

> . . . the Alpine battery — a standing example of the reckless style unfortunately far too common in Otago of placing valuable and expensive machinery on untried claims. Although this is one of the best reducing plants of its class being a ten-head battery driven by an overshot wheel of iron I should think by appearances it has never crushed fifty tons of stone. This battery was erected at enormous expense . . . The reef was found, however, not to contain gold in payable quantities.[5]

Figure 13. 3. Looking up Battery Creek. The long ridge running across the photograph (with the jagged Garvie Mountains behind) divides the catchment of the Clutha River from that of the Mataura River. The approximate site of the Alpine Quartz Reef Co. battery and water wheel is arrowed.

A later comment was somewhat kinder:

> It was abandoned owing to its inaccessible situation on the mountain and the consequent high cost of everything brought to the mine, coupled with the high costs of wages ruling in those days and the primitive methods of saving gold in comparison with the present up-to-date machinery and cyanide.[6]

The battery was bought by Henry Symes in February 1896 with the intention of shifting it down to his Exhibition Mine on the front face of the Old Man Range above Bald Hill Flat. However, for some reason this plan did not go ahead and the dismantled remains of the battery still lie in remote Battery Creek.

The big water wheel was dismantled during the summers of 1968-70 by a group of enthusiastic and hardy committee members of Alexandra's William Bodkin Museum. The pieces, which totalled an estimated five tons in weight, were carried out of the steep gully by tractor and helicopter and then transported to Alexandra by truck where they were re-erected outside the Museum in Thompson Street.[7]

Figure 13.4. The water wheel at the Alpine Quartz Reef Co. claim with remains of the battery to the' left. The storage dam for the Fraser Basin Hydraulic Elevating Co is arrowed. The summit ridge of the Old Man Range forms the skyline with the Obelisk (1,695 m) to the left.

Fraser River Hydraulic Sluicing Co

Charlie Nicholson's faith in the Fraser River as a gold producer was apparently not discouraged by the failure of the Alpine Quartz Reef Company. In 1890 he became interested in mining the terraces along the river, more than a mile below the junction of Battery Creek in which the abandoned battery and water wheel were located. He formed the Fraser River Hydraulic Sluicing Company, again made up mainly of friends and members of his family, with Thomas Scott as mine manager. A water race to the claim, from a point further up the Fraser River, took a long time to build, and when it was at last finished, it was found that the water supply was erratic and insufficient.

Figure 13. 5. The water wheel of the Alpine Quartz Reef Company erected in front of the William Bodkin Museum in Alexandra.

About five years went by before any further work was done on the claim. Because of the rigorous climate, working time was limited, but there were other problems too. The gold-bearing wash lay beneath a thick layer of gravel with interbedded clay, a mixture very difficult to deal with. Furthermore the wash was far below the water table and the mine could not be drained. It was clear that hydraulic elevating was the answer, but to raise the money required, it was necessary to form a public company.

Fraser Basin Hydraulic Elevating Co Ltd

The company was floated in early 1899 with a capital of £2,100 and 35 shareholders, mainly from the Lawrence district.[8] The claim was enlarged to 60 acres, a dam constructed in Battery Creek and a high-level race cut to a point high above the claim. Pipes were carted in to convey the water from the end of the race to the elevator. In spite of all this effort, gold was sparse and difficult to recover. The end came in November 1904 when the company went into liquidation.

Hopes lingered on however, and another attempt was made in 1906 when 'a strong local syndicate' (Duntulm Hydraulic Sluicing Party), which turned out to be again led by the Nicnolsons, took up a claim in the head of Fraser Basin. The elevator used by the defunct Bald Hill Sluicing Company was laboriously carted into the basin before the winter

Figure 13.6. The Fraser River Hydraulic Sluicing Company's claim, with worked ground to the right in the bed of the Fraser River. Its water race from the Fraser River is below the hut. Later the Fraser Basin Hydraulic Elevating Co Ltd brought in the high level race from Battery Creek.

of 1907, and in January 1908 another start was made. But this attempt met with no better success than previous ones.

Even as late as 1915 a member of the family was apparently still pottering about in the Basin, as an application by William C. Nicholson for an Extended Alluvial Claim of 3 acres, is recorded.

The 'Sisters' Diggings.'

Cradling and ground sluicing, sometimes accompanied by diversion of the river by flood race or by wing damming, took place along most of the course of the Fraser River from the outlet of the swampy basin at the head, to the Fraser Dam basin. Tributaries such as Rough Creek and Frenchmans Creek were worked also. Generally the amount of workable ground was in small patches, but apparently supported a large, fairly settled population of miners, as Bristow records no fewer than 57 living sites along the river. But it was in the broader valleys of the northern part of the Basin with low terraces and wider flood plains that allowed the most extensive ground sluicing.

It is likely, from their extent and longevity, that these diggings, named the 'Sisters diggings,'[9] were the most successful of any on the river.

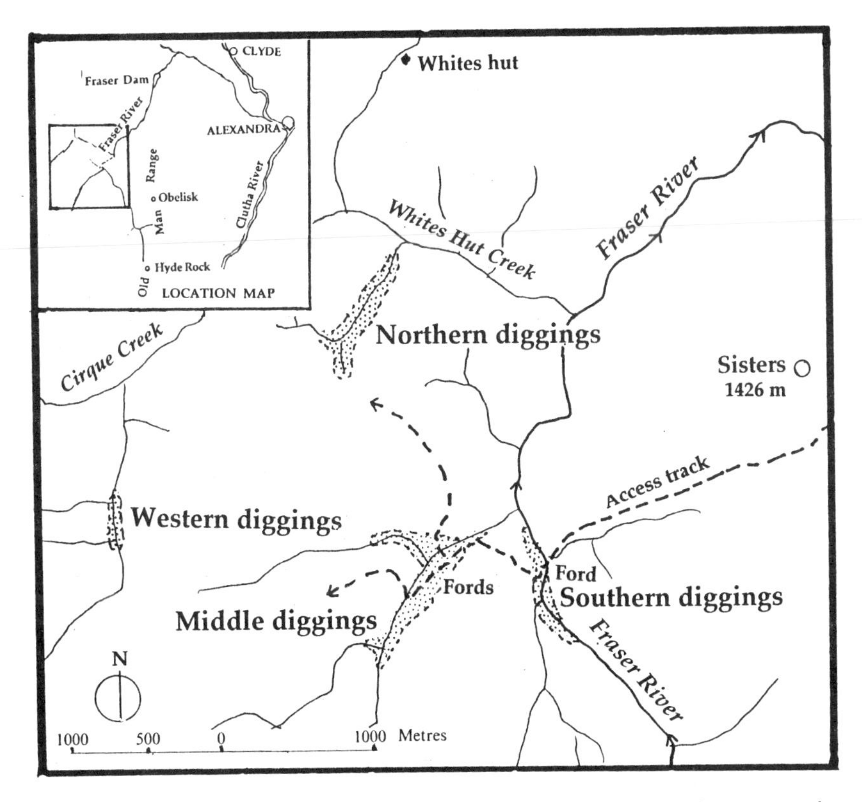

Figure 13. 7. Gold workings on the Fraser River and its tributaries, named collectively 'The Sisters Field' (after the nearby Sisters rocks) by archaeologist, Dr Jill Hamel who also designated the separate diggings by their geographical position as shown on the map.

Within about two kilometres of the ford where the Sisters 'road' crosses the Fraser River lie at least four areas of ground-sluicing on the river and its tributaries. These have been named by Dr Hamel, for convenience of description, as the Western, Northern, Middle and Southern Sisters diggings.

HAWKS BURN BASIN

In the small basin that lies upstream from the junction of the Fraser River with its major tributary, the Hawks Burn, is a narrow, low-lying alluvial flood plain of small extent. On the southern side, a remnant of a higher gravel terrace lies close under the very steep schist wall that marks the southern boundary of the basin. A small outcrop of the old Tertiary sediments that once covered much of the landscape is preserved under the terrace gravels, and indicates that the little basin probably originated as another of the down-faulted depressions that are so common in Central Otago.

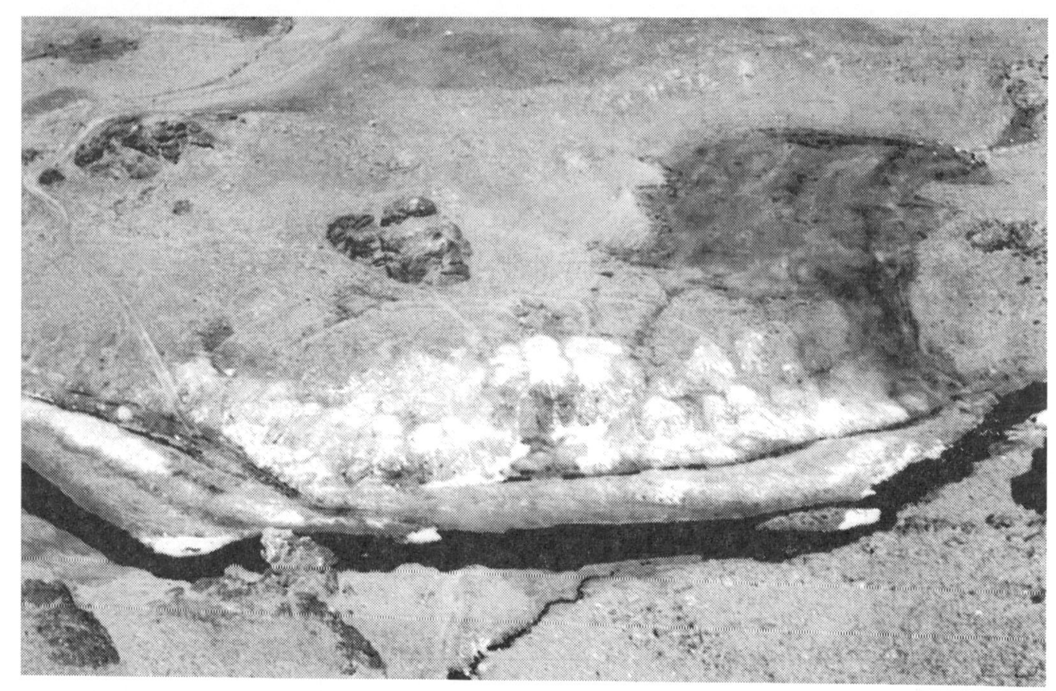

Figure 13. 8. Part of the Southern Sisters diggings on the Fraser River also shows a section of the one kilometre-long diversion channel.

Figure 13. 9. The Middle Sisters diggings. Ground sluicing at the second ford where the access track crosses a substantial tributary of the Fraser River.

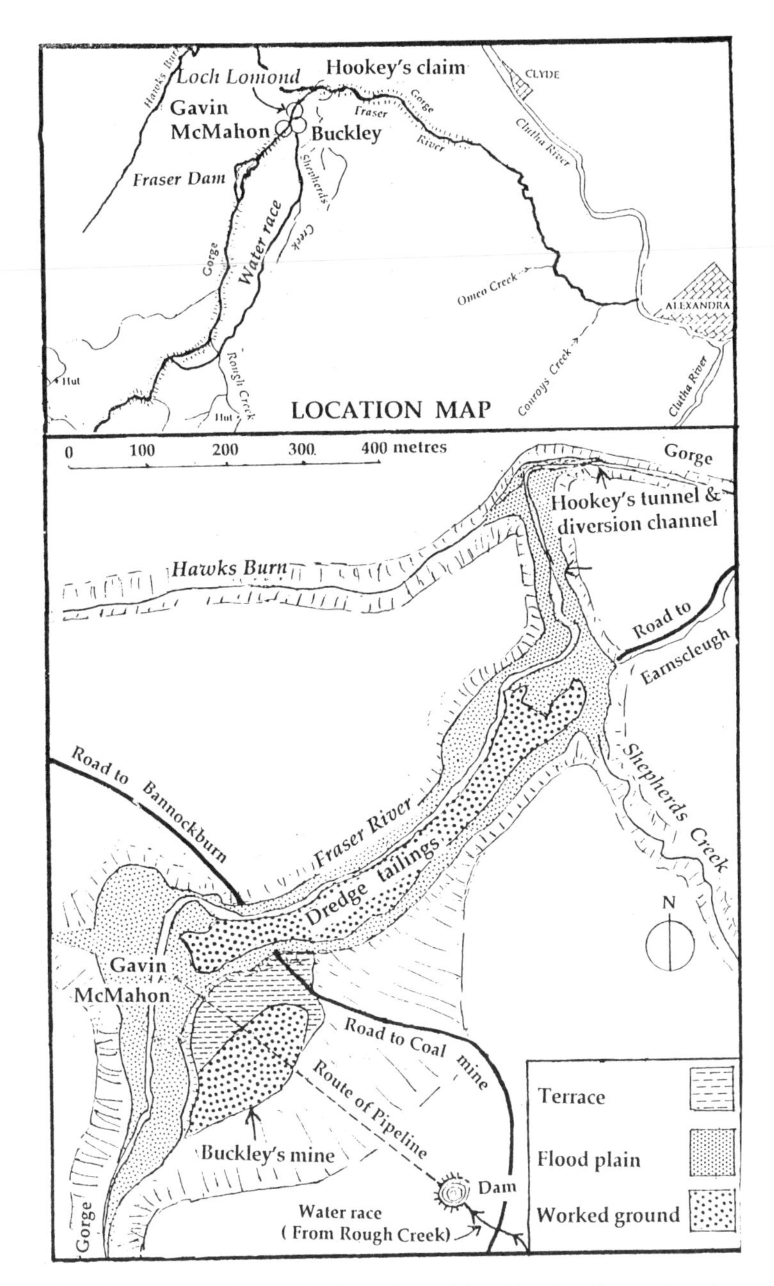

Figure 13. 10. The basin above the junction of the Hawks Burn showing principal Gold claims as well as other features.

In 1886 a party of miners led by James Gavin, and joined later by Owen McMahon, planned to mine the flood plain, reasoning that the basin, lying as it does at the foot of a steep gorge, would be a natural trap for gold. They constructed a race about five miles (8 km) long from Rough

Creek, leading the water by way of Shepherds Creek, where a storage dam was constructed, into a 'C'-dam* overlooking the claim they had taken out in the river bed. The plan was to mine by means of an hydraulic elevator, but they had far too little water, so they had to extend the race from Rough Creek into the gorge of the Fraser River

Mining had started by 1893 and 75 oz of gold had been recovered, when a flood swept a mass of tailings from the workings down the Fraser River and deposited it on the paddocks of Earnscleugh Station down on Earnscleugh Flat.[10] The annoyed runholder, Laidlaw, took out an injunction and so put a stop to the work.

Meanwhile John Buckley and partners had taken out a claim up on the higher terrace. They took over McMahon's water supply but they did not need to use an hydraulic elevator as they had sufficient fall to dispose of their tailings. They mined by hydraulic sluicing.

The Dredge

Although the 'dredging boom'— that mad speculation in dredging shares — was over by 1903, gold dredging was still the fashion of the hour. New dredges were being planned, built or launched every other week. No stream was safe from the attention of the steel monsters that were rapidly replacing the small and relatively inefficient machines of the 1890s.

Figure 13.11. Extensive ground sluicing on the terrace adjoining the Fraser River in the Hawks Burn basin was carried out by John Buckley and his party.

Apparently as early as June 1900, it became known that a group of Cromwell investors were planning to dredge the alluvial flat immediately upstream from the junction of the Hawks Burn and Fraser River. This information greatly alarmed William Hookey who, after a prodigious effort cutting a tail race, was about to start sluicing the lower end of the same flat.[11]

The Cromwell group, now calling itself the 'Loch Lomond Company,' turned its attention to the Shepherds Creek dredge that had been dredging an 89 acre claim on the Bannock Burn. This dredge had started work in February 1901, but only three months later the company was in liquidation and the dredge was up for sale. After standing idle for nearly a year, the Loch Lomond Company bought the vessel, and a tender of £550 from Knewstubb was accepted for dismantling and re-erecting it. Although the dismantling was completed in June 1903, it was too late in the season to cart the large amount of machinery over the saddle to the Fraser River. McLoughan, who had the contract, had only commenced the cartage when frost and snow brought things to a halt.

It wasn't until October that the pontoon was launched on the new site, but after that the erection of the dredge went ahead quickly under the supervision of Dredgemaster W. Sanders and Engineer Maurice McCraw, so that it was able to begin work in December. The dredge was built on a wooden pontoon 70ft (21 m) long, 26 ft (8 m) wide and 5 ft (1.5 m) deep. It could dredge to a depth of 6 metres with buckets that held 3.75 cubic feet (0.1 cubic metres) of material, and discharged the tailings by means of an elevator 45ft (13.7 m) in length.[12]

The dredge began well, bringing up gold from a depth of 16 feet, but not without difficulty. Because the basin lay at the foot of a very steep section of the gorge, it was the repository not only for gold, but also for large boulders that had washed down the river. The rough work began to take its toll and it was soon apparent that the dredge was too lightly constructed for the conditions.

First, it was the screen that was smashed, but it was repaired in the Alexandra foundry. A month later there was much more serious trouble when the string of buckets came off the bottom tumbler, the tumbler broke and so did the bucket line which promptly fell into the dredge pond. The dredge master was replaced but this did not stop the constant breakages.

Supplying coal to the dredge in such an isolated place would have been a major problem had not coal been discovered high on a nearby ridge and less than three kilometres from the dredge. James Holt of Clyde began carting coal along what appeared to be a reasonably good road, and it was hoped that he would ensure a regular supply. But it turned out that the road was difficult in the winter and coal supplies became erratic. Holt was blamed, so the dredge company applied for a coal licence itself, and in October 1904 it bought Holt's mine.

Coal supply was not the only problem caused by the winter. Whereas most dredges on the Clutha River welcomed the winter because the water level went down and allowed the dredge buckets to reach the bottom more easily, the cold weather caused great problems to the *Loch Lomond*. It was not just the cutting up of the coal road that caused trouble. Horses pulling the coal wagons actually died from the cold. Furthermore the winter temperatures were so low that the men could not live and work under the conditions. The dredge had to close down for several months each winter.

Returns of gold were erratic, ranging from 5 oz to 25 oz a week. Unfortunately the lower figures were more frequent, and as it required at least 15 oz a week to cover expenses, the dredge was not very profitable. For the year 1905, for example, the dredge worked six acres of ground for 609 oz of gold, worth £2,359, but the expenses for the year came to almost exactly the same figure. Stoppages caused by damaged gear and closing down during the winter, meant that only 34 weeks dredging had been accomplished for the year.[12]

Now another problem was coming up. The dredge had worked its way up to almost the top of its claim, where the ground was exceedingly

Figure 13. 12. The dam at the end of the water race from Rough Creek. The water worked the hydraulic elevator set up by McMahon in the bed of the Fraser River a kilometre above the Hawks Burn junction.

bouldery, and in early 1906 the bucket ladder and the shaft of the bottom tumbler were bent. Then there was trouble with timbering in the coal mine, which stopped production and brought the dredge to a halt. It had only started again when the breaking of a bearing in the main crown wheel stopped it for another two weeks.

A month before the dredge reached the top of the claim, the company bought the claim of William Hookey adjacent to the lower end of its own claim, but it would be necessary to wait for the spring floods before attempting to take the dredge the kilometre downstream to the new claim. Meanwhile the dredge was lying idle. It was bought [13] in September 1907 by a syndicate of miners from the Nevis Valley who spent the next month, under Dredgemaster Ted McDonald, overhauling the vessel as she had become silted up in her pond. When the dredge was back in good order, they turned her round with the idea that she would dredge her way down through her own tailings to the new ground, but it was not to be. Turning round was as far as she got, and she then lay idle for several years.

In June 1911 the dredge was bought by the newly formed Carrick Gold Mining Company that wanted the boiler and engine to work its quartz mine on the Carrick Range above Bannockburn. Iron from the superstructure and timber from the pontoon were used for mine buildings.

Figure 13.13. The *Shepherds Creek* dredge before it was dismantled and re-erected in the Hawks Burn basin as the *Loch Lomond.*

Ten years later, when the machinery of the Carrick mine was sent off for scrap, the last remnants of the *Loch Lomond* dredge disappeared with it.

Hookey's Dream

One of the most unusual and innovative mining efforts took place at the entrance to the gorge that leads away from the basin that lies above the junction of the Hawks Burn. It was a scheme that involved a huge amount of time-consuming physical work — the sort of enterprise usually undertaken by large companies, not by three men working part time.

Bill Hookey[14] and his mates, however, were not afraid of hard work. Bill was an old-time miner who had worked at the Serpentine Diggings, at Blacks and on the Old Man Range. His mates, James McConnell and James Chalmers, came from the Nevis. Their method of financing the work was to go off shearing during the spring and summer and work on the river during the autumn and winter.

The group's first application, complete with sketch (see next Chapter), was for a claim in the Fraser River gorge below the Hawks Burn junction. It was approved in March, 1899. This scheme involved a short flood race into which the river could be diverted, and an open tailrace in the de-watered river bed.

Figure 13. 14. A heap of dredge tailings marks the spot where the *Loch Lomond* dredge commenced operations in 1903. It was here that Hookey's diversion wall was rendered ineffective by the dredge.

However, it was then decided that the riverbed in the basin above the Hawks Burn junction was a better prospect, so another application was submitted two months later. This was a more ambitious scheme and required much more preparation before actual mining could begin.

Bill Hookey was certain that gold lay under the alluvium in the basin, but it was too deep to be reached by conventional means. So the party decided to begin some distance down the steep gorge and drive a tunnel under the length of the river until they reached the alluvial flat. The tunnel would act as a tailrace for getting rid of the tailings as they sluiced away the gravel to expose the gold lying on the rock bottom.

Their application to the Warden's Court not only asked for an Extended Alluvial Claim of three acres (1.2 h), but also included a flood race into which the river could be diverted while a tail race was being constructed.

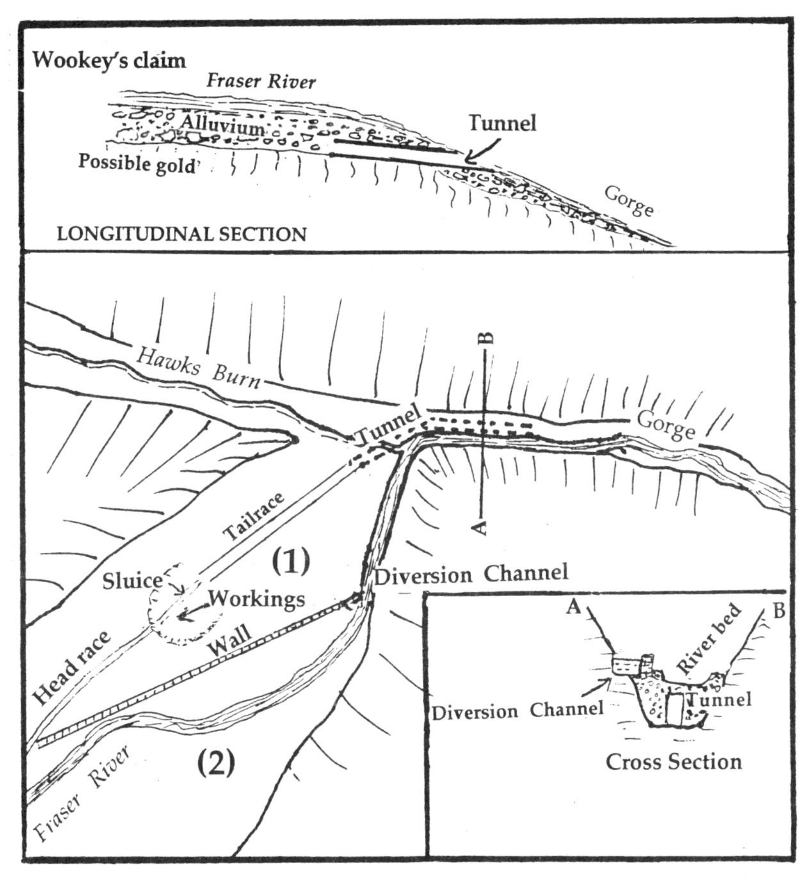

Figure 13. 15. Diagrams to illustrate Hookey's plans for mining the basin at the Hawks Burn junction. The Fraser River was diverted into the Diversion channel exposing Area (1) of the river bed for working by ground-sluicing, When Area (1) was worked out the plan was to divert the river to the other side of the wall so Area (2) could be worked.

Figure 13.16. It was in this gorge that Hookey and his mates laboured for four years to prepare for sluicing in the Hawks Burn basin only to have their work ruined by the *Loch Lomond* dredge.

A separate application was made for the all-important tailrace tunnel. The applications were granted in July 1899.

First, the large flood race, capable of carrying the normal river flow, was cut along the side of the gorge. This was a major work in itself, as it was 20 chains (400 m) long and up to six feet wide with stone walls in places more than six feet high. Much of it was cut through solid rock, and large quantities of explosives had to be used. Several times floods washed the walls away, and one bad break took all of a winter to repair. Little wonder it was two years before the diversion channel was completed and the river turned into it. Once the riverbed was dry, they began to construct the tunnel under the huge amounts of coarse alluvium in the mouth of the gorge.

Some of the boulders under which they tunnelled were estimated to weigh 20 tons, but after 100 yards they were through the worst of the debris and a different approach was adopted. The tunnel, which averaged about five feet (1.5 m) high, was extended by excavating an open trench and then covering it with large slabs of rock dragged from the hillsides, and lifted into place with a homemade 'jinny.'[15] Finally by August 1902 the 10 chain (200 m) tailrace tunnel was complete, but there was still more work to be done before they could start mining.[16]

A wall had to be built up the middle of the flat to confine the river to one side while they ground-sluiced the other side. This was really a type of wing-damming and the work was carried out over another year. The amount of water they required for sluicing was to be taken from the river and then led down the tunnel, while the bulk of the river water would be diverted, first, behind the wall, and then finally into the big flood race. The partners had completed the diversion wall by the end of the winter of 1903 and were looking forward to commencing mining in earnest when they returned from their summer shearing.

Imagine their horror and disappointment when they found that the *Loch Lomond* dredge, which had been erected and commenced work while they had been away, had dredged down both sides of their wing wall, effectively depriving the party of any return for their years of hard work.[17]

NOTES

1. *Otago Daily Times* 20 November 1862.
2. In Grey 1927 pp 575-78.
3. Letter from Robinson in Pyke, 1887 p. 102
4. NZ Cyclopaedia.
5. *Dunstan Times* 6 June 1890.
6. *Otago Witness* 4 June 1896.
7. Hannah, M. n.d. (c1975).
8. Company Records National Archives, Dunedin Regional Office. Although Francis Nichols, a mine manager of Lawrence, held 170 shares four members of the Nicholson family James Carle, William C, Charles George and Samuel held 120 shares each.
9. Named after 'The Sisters,' a prominent group of hilltop rocks lying to the north-east, by Dr Jill Hamel who carried out an archaeological survey of the Fraser Basin in 1989-90. (Hamel 1994)
10. *Alexandra Herald* 4 December 1901
11. Galvin, P. Mining Handbook 1906.
12. *Ibid.*
13. *Dunstan Times* 28 September 1907
14. In the next chapter it is shown that Bill Hookey's name was actually William Wookey, but as all his applications for mining privileges on the Fraser River were applied for under the name 'Wm Hookey,' this name is retained in this chapter.
15. A simple type of crane. Almost certainly a corruption of 'jenny' which was a name applied to a mobile railway crane and also to the big wheel at the top of a 'poppet-head' over a mining shaft.
16. *Otago Witness* 4 December 1901.
17. This information about the dredging of Hookey's claim by the *Loch Lomond* is related by Syd Stevens (1988). Stevens heard the story from Hookey himself when both were in Dunstan Hospital in the mid-1930s.

14.

WHAT'S IN A NAME?

— Mainly about Bill Wookey

One of the first problems encountered in investigating the history of the goldfields of Central Otago, is that of names. First, there is the difficulty of identifying landscape features from the names given to them by miners, as many do not appear on present-day maps. This has already been mentioned, particularly in regard to the diggings at Waikerikeri Valley, and along the West Bank of the Molyneux. These miners' names were freely used on official documents such as water race licences, and were accepted in the Wardens Court, but it often requires a good deal of cross checking to identify the features referred to. Secondly, there is a wide variety of spellings used for peoples' surnames, particularly in newspapers, but also in official documents.

Most of the information about Central Otago published in the Dunedin newspapers was garnered by 'special correspondents': people in the upcountry towns who undertook to write about local affairs, and were presumably paid on the amount of their material that was published. They invariably wrote their reports in longhand and the typesetters at the newspaper office had to translate the correspondents' writing. By and large the typesetters, who were experienced at reading difficult handwriting, did very well with normal English. They did what everyone reading difficult handwriting does—filled in the bits they couldn't read.

When it came to surnames though, it was a different matter. The name of Jeremiah Drummey, the bridge builder and contractor, for instance, seemed to give the typesetters particular difficulty. His name appears as Drummond, Drummy and Drumey as well as Drummey, in various articles. Another surname that gave trouble appeared in the Electoral Roll of 1880-81 as 'Slevin,' but as 'Sleven' in 1884-85. In the Burial Register of Clyde Cemetery it is 'Slaven,' but on the headstone it is spelt 'Slevin.'

Not all of the difficulties were caused by hand writing. Some miners were barely literate so needed help in filling in forms, and by the 1890s many were now old men, but still had marked provincial or foreign accents. The difficulties of a clerk of the Warden's Court in trying to understand the answer when asking such a person for his name can be imagined. The clerk would write down what he thought he heard, and that version would then be part of the official record.

Man of Many Names

In the last chapter we met Bill Hookey, the man who, with a couple of mates, undertook what was a big task for the time, of diverting the Fraser River into a flood race. His applications to the Wardens Court and other correspondence were all signed 'William Hookey,' as in a classic letter to his solicitor reproduced below. Bill had run up against new regulations

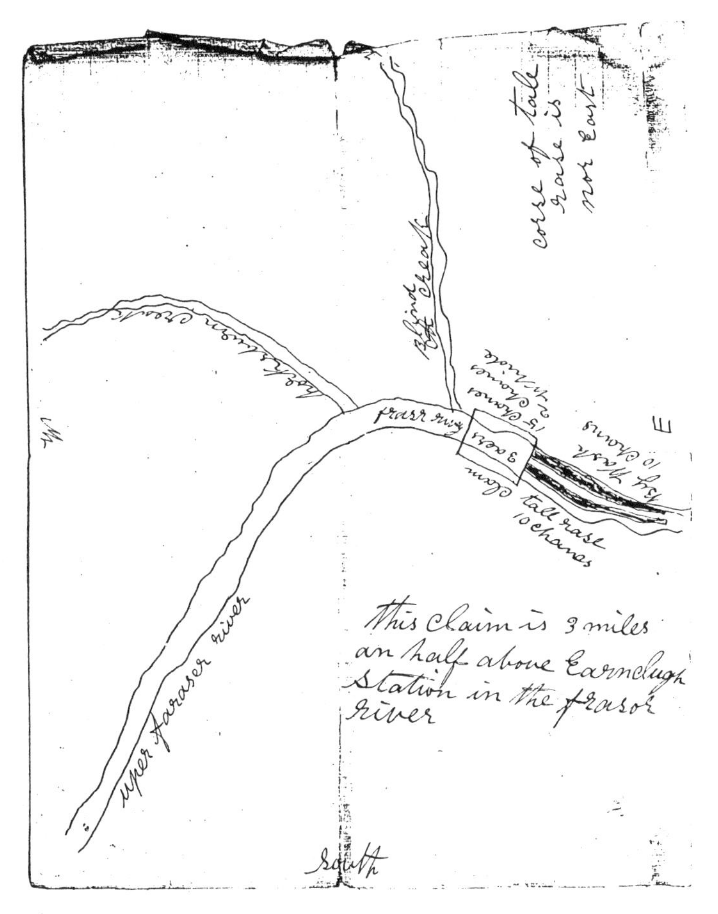

Figure 14.1. Sketch submitted by Wm Hookey (Wm Wookey) to accompany his first Application for a claim on the Fraser River. Note the odd spelling. Another application was made two months later for a claim above the Hawks Burn junction (see Figure 13.15).

that required him to supply a sketch plan of his proposed claim with his applications. He was not very pleased when the solicitor Robert Gilkison of Clyde, acting on his behalf, asked him to provide the required sketch.

It is no longer possible to copy Hookey's original letter to Gilkison in his large, but now faint, handwriting but an exact reproduction is:

> Feb 27/1899 Kawarau Station
> Mr Gilkinson Dr Sr
> i foward your plan of mye claim as near as posibl i dont profest to be an artist this new mining rool is complet hombug and it means to stoop mininige and prospty to much troublr to Get claim Granted an a man want to be well up in lor i object to this roole for the reason a man Want to Employ artist to sketh claim & want to employad a saw mill to saw 30 feet of timber to mark claim With i think little of the mining rools iff the claim is granted please Lett me know as soon as poseble
> Yours truly
> Wm Hookey

Translated, this reads:

> Mr Gilkison, Dear Sir,
> I forward your plan of my claim as near as possible. I don't profess to be an artist. This new mining rule is complete humbug, and it means to stop mining and prosperity. Too much trouble to get a claim granted and a man needs to be well up in law. I object to this rule for the reason that a man needs to employ an artist to sketch a claim and would need to employ a saw mill to saw 30 feet of timber to mark the claim with. I think little of the mining rules. If the claim is granted please let me know as soon as possible.
> Yours truly

In Hospital

We next run into Bill Hookey in Dunstan Hospital during the mid-1930s. His ward mate was Syd Stevens who included a story about Bill in a book of Central Otago tales he wrote in the 1980s.[1] At the time Syd knew him, Bill was a very old, sick man who had had one leg amputated because of gangrene and was suffering from bronchitis and other complaints. He was obviously going to end his days in the hospital. Syd tells how Bill regaled him with oft-repeated tales of gold mining, ending each session with an invitation to Syd, to join him in an expedition into the hills. "I know a place," he would say "Where we can pick it up by the handful. Just wait till we get out of this hospital." But he never did get out.

The Central Otago poet, Tod Symons, read Syd Stevens' story and wrote a poem about an old miner, 'Bill Whookey', who was obsessed with getting back to find the gold just waiting to be picked up. *I Know a Place* was the title of Tod's poem.[2]

In the same decade that Syd Stevens was in Dunstan Hospital, another writer was there. The Southland historian and journalist, Fred Miller, himself a temporary gold miner at the time, subsequently wrote a story for the *Otago Daily Times*.[3] He told how he had just spent some time in Dunstan Hospital with a poisoned foot and along with other patients, had listened each day to the tales of an old miner in the hospital who had had a leg amputated because of gangrene. He told tales about gold back in the hills just waiting to be picked up. Fred Miller called him 'Bill Wookey.'

William Wookey was a well known miner on the Old Man Range above Bald Hill Flat (now called Fruitlands) who had, along with Gavin, in 1882-83 opened up the Exhibition Reef as it was later called, (see Chapter 6, Figure 6. 3, Section 23) which produced reasonably good returns for a time and then petered out. He also had an interest in other claims on the slopes of the range.

It seems an extraordinary coincidence that there should be two men with nearly the same name telling the same kinds of tales in Dunstan Hospital at the same time. It appears that not only were 'Hookey' and 'Wookey' one and the same person, but that Wookey's name could be spelt in a variety of ways.

The hospital's Patient Register[4] showed that a 'William Whokey' had been admitted in October 1931, and had had his right leg amputated because of gangrene, two months later. After a stay of 1,664 days, he had died in the hospital on 7 May 1936. The hospital thought the patient was aged 92 at the time of his death. It also thought that he was of Australian nationality, although his Death Certificate says he was born at sea.

It is interesting to note that the hospital charged each patient 21 shillings a day, so Bill's stay cost £998-8 shillings, but as he was an Old Age Pensioner he was charged only two shillings a day. Even this was too much apparently, and Bill died leaving a debt to the hospital of £166-8 shillings.

Fred Miller had also mentioned in his article that before being admitted to hospital, Bill Wookey had lived in a little hut on the bank of the Clutha River just above the Alexandra bridge. Sure enough, a search unearthed a mining map, dated 1926, which showed a claim labelled with yet another spelling variation. Just above the Alexandra Bridge on the Earnscleugh side is a section clearly labelled 'W. Wookay's claim.'

Were Bill Hookey, Bill Whookey, Bill Wookey, William Wookay and William Whokey all the same person? And if so what really was his name? A search of the Register of Deaths showed that only one Wookey had died between the years 1926 and 1946 and this was William Wookey who died in 1936. The only Hookey who died in those same 20 years was an Alfred Tily Hookey who died in 1936. The Death Certificate of William Wookey confirmed that he had died at Clyde in 1936 and the Clyde Cemetery records list a headstone inscribed simply 'Wookey.'

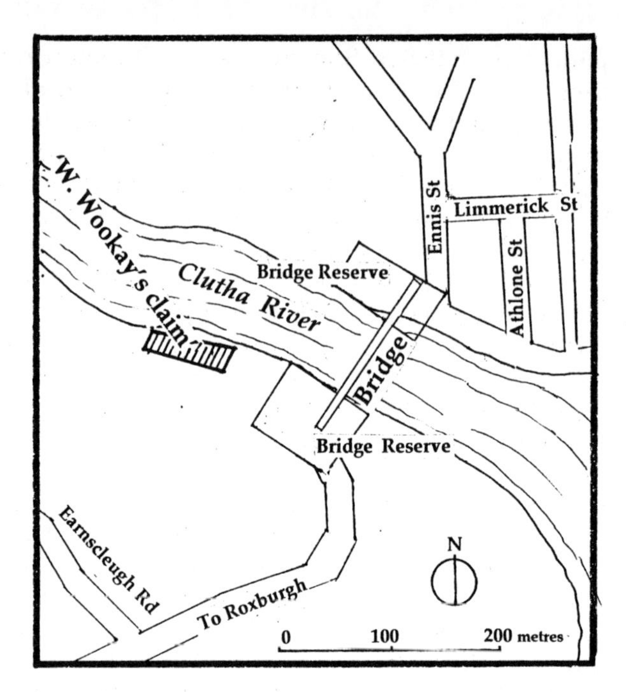

Figure 14. 2. Location of the claim of 'W. Wookay' above the Alexandra bridge on the southern side of the river. Adapted from Mining Plan S. 0.860, 1926.

The was no death notice in the paper, presumably because Wookey had no relatives or friends to insert such a notice, but the final chapter in this little story was a short obituary published in the *Dunstan Times.*[4] It mentions that Bill Wookey had been a miner in Central Otago since the earliest days and each summer had worked as a shearer. It goes on to say that Bill Wookey was also known as Bill Hookey. It may be assumed that the obituary was written by his ward mate and gold mining partner-to-be, Syd Stevens.

It is also evident that all these references are to the same man, William Wookey.

Notes

1. Stevens S. 1988 pp. 30-31.
2. Symons, Tod 1978. pp. 23-24.
3. *Otago Daily Times* 5 January 1933.
4. Held in the Dunedin Regional Office of National Archives.
5. *Dunstan Times* 11 May 1936.

15.
'COOLGARDIE.'
—Rush to the Manor Burn

The excitement in Alexandra reminded old timers of the earliest days of the Dunstan Rush. But this was 32 years later — Sunday 15 July, 1894. Two strangers, Walter Murray and Harry McDonald, had come into town and announced that they had recovered 20 oz of gold for a week's work from the 'Manorburn terraces,' about two and a half miles (4 km) up the Manuherikia River from Alexandra.

The news flew round the town, and according to the newspaper report of the time:

> . . . nearly the whole population of Alexandra flocked out to the ground, and claims were pegged out right and left. Some parties on the spur of the moment pegged out claims without holding miners' rights and rushed away to the township to take out rights but on coming back found to their sorrow that their claims had been jumped.[1]

Others, to make sure that they held their ground, slept in the open on the claims they had selected. Within a few days 30 claims had been pegged out and the next session of the Warden's Court was busy hearing pleas for 'Extended claims.' The size of the claim awarded depended on the number of miners in the party, and was generally at the rate of one acre for each man. So Lachlan Campbell and William Fawcett, for example, were granted one acre each, and two-man parties such as those led by Samuel Jackson, John Bruce, Edwin Appleton and Robert Kinnaird were granted two-acre claims. The prospectors who discovered the gold were granted a double-sized claim together with permission for a dam in a nearby gully, and a water race from it to their claim. A larger party headed by Lewis Cameron and John Pattison was granted a seven-acre claim next to the prospectors, but within a week they had given it up. In fact it seemed that the only claim to produce any gold was that of the prospectors.

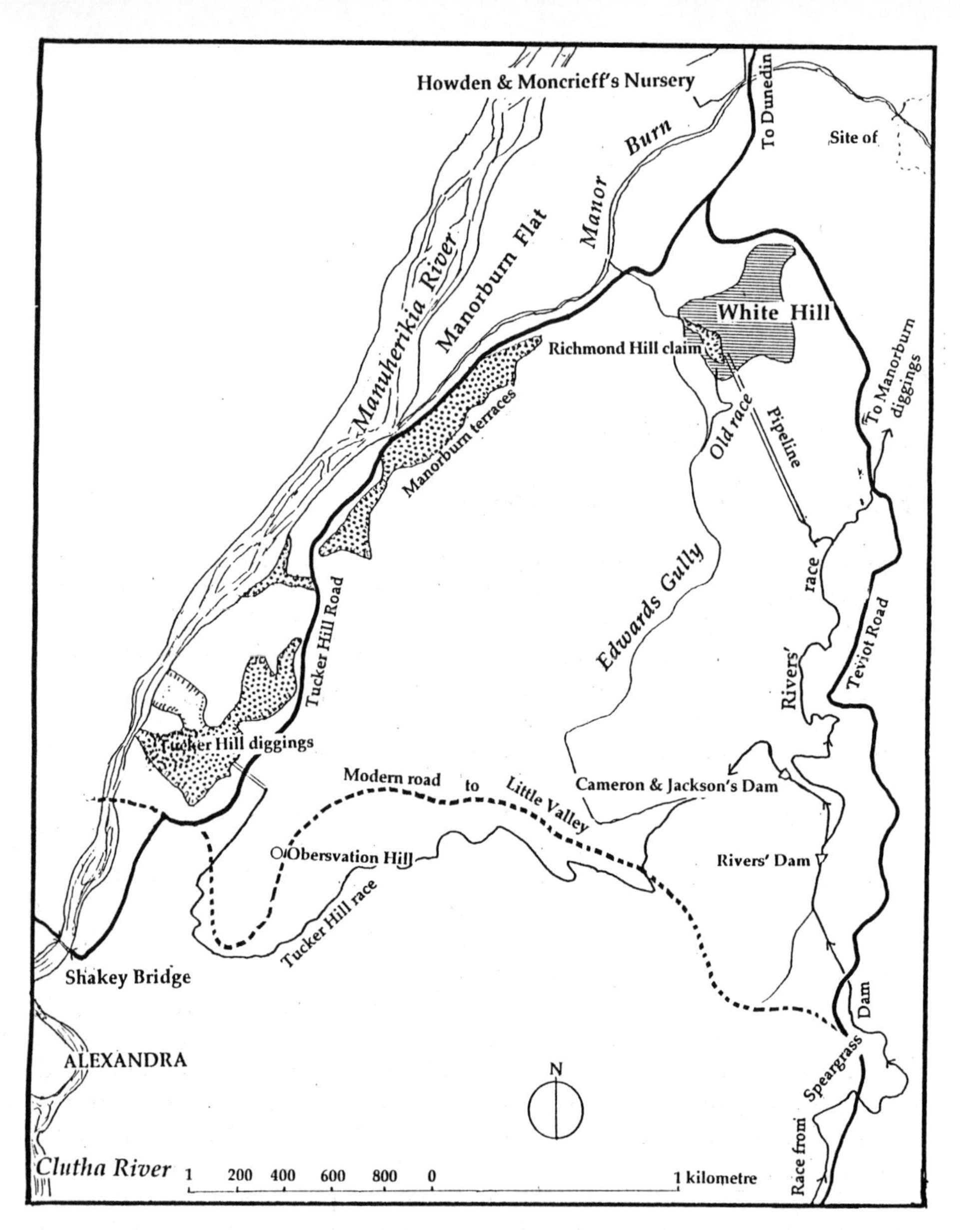

Figure 15. 1. Location map showing White Hill. Also shown are the water races Supplying Tucker Hill diggings and Rivers' Richmond Hill claim, and other features.

The Manor Burn

The Manor Burn,[2] a substantial tributary of the Manuherikia River, drains the rugged rocky country to the south-east, and flows from a narrow, steep rocky gorge out on to the flood plain of the river at the lower end of

Galloway Flat. Instead of joining the river directly, however, the Manor Burn turns at right angles and flows parallel to the larger river for nearly two kilometres before merging about a kilometre above the Alexandra town boundary. The delta-shaped, low-lying ground between the two streams was known as Manorburn Flat, and was the scene of intense gold seeking activity almost from the beginning of the Dunstan Rush in August 1862.

Alluvial terraces flank the southeastern bank of the Manuherikia River (with a few short breaks) from its junction with the Clutha River right up to the entrance to the Manuherikia Gorge, about 12 kilometres upstream. The kilometre or so of terraces flanking the lower course of the Manor Burn were known as the 'Manorburn terraces' although it is likely that they were actually formed by the Manuherikia River. Although these terraces did contain gold and had been mined at various times over the years, they were not the source of the present gold discovery.

White Hill

Murray and McDonald had made their find at White Hill, a low ridge standing nearly a kilometre back from the river and about 60 metres above it. Miners recognised it as a 'made hill,' meaning that it was not formed from solid rock, as were most of the ridges in the vicinity, but from

Figure 15.2. The remains of White Hill are now a highly visible landmark because of the excavation of sand for building. This modem view looks across the willow-filled course of the Manuherikia River from above the mouth of Letts Gully.

alluvium or other sediments. It was in fact composed of white or yellowish quartz grit and gravel (the 'granite wash' of the miners) overlain by a thin layer of schist alluvium. The hill covered an area of about 75 acres (30 h).

The two prospectors had enlarged and deepened a shallow hole dug by some miner years before. After sinking three feet or so, the old timer had struck white quartz gravel, and mistakenly thinking that he had reached the 'bottom,' he abandoned the excavation. Had he gone down a little further he would have struck the gold-bearing wash that the latest prospectors had discovered.

Within three weeks of their discovery, the prospectors, who seemed anxious to move on, sold out to William Pacey and James Nieper, and left the district. They apparently did not know, until some time after the sale, that much richer gold than they had found had lain only a short distance below their feet.

The mine taken over by Pacey and Nieper consisted of an excavation measuring 10 by 14 feet (3 by 4 m) in the side of a spur about 30 feet (9 m) above the floor of Edwards Gully,[3] a small dry valley leading down to the Manor Burn. The back wall of the excavation, about 10 feet (3 m) high, was in the whitish granite wash. Below the granite wash was 'blue clay' and it was on this that the prospectors had obtained their gold. Six hours cradling of this material produced gold to the value of £6. The miners assumed that this blue clay was the 'bottom.'

Figure 15. 4. Edwards Gully with the water race probably built by Pacey and Nieper in anticipation of obtaining water from Campbell and Jackson's race further up the gully.

The neighbouring claim, recently abandoned by Cameron and party after sinking a shaft about 14 feet (4 m) deep, had been taken over by J. E. Thompson and Wm Noble who deepened the shaft in an endeavour to reach the gold bearing wash exposed in Pacey's claim. They did better. They penetrated through the 'blue clay,' and at a depth of 16 feet (5 m), found another layer of rich wash resting on the true bottom of decomposed schist. The discovery that the prospectors had obtained their gold on a false bottom and that much richer wash lay below put a different complexion on the value and longevity of the field.

Owing to the rugged, rocky terrain, it was too difficult for Pacey and Nieper to bring in a water race from the Manor Burn at the height necessary to sluice their claim. So they tried to arrange a supply from Campbell and Jackson, whose race on its way to their claim at Tucker Hill, crossed the head of Edwards Gully. However, the arrangements fell through, but not before Pacey and party had constructed a water race leading out of Edwards Gully, which was, presumably, intended to convey the water to the White Hill claim. This short race, supported by well-built stone walls, still exists today.

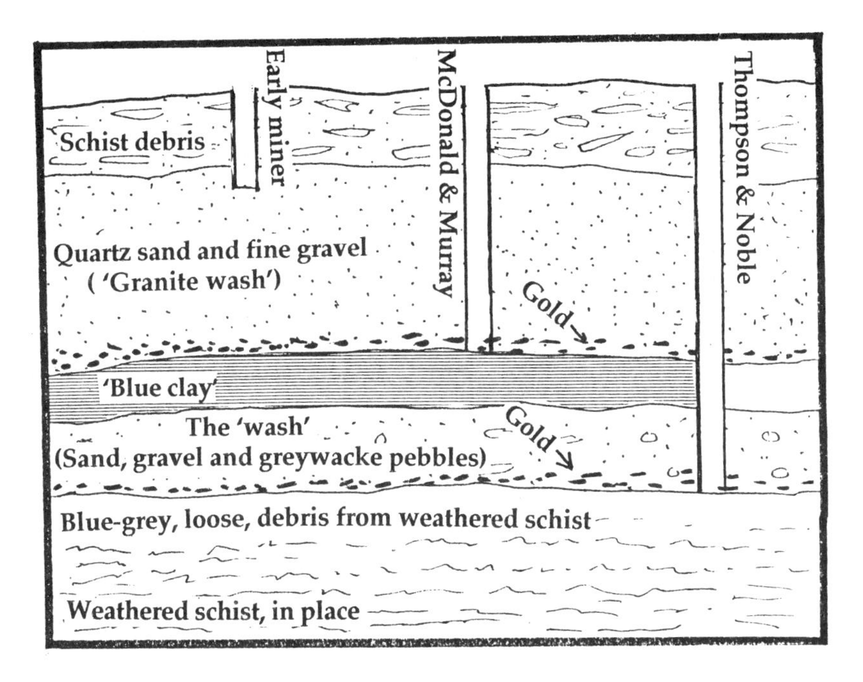

Figure 15.3. Diagrammatic cross section of White Hill showing strata and shafts sunk by prospectors. An early prospector thought he had struck 'bottom' on 'granite wash.' McDonald and Murray struck gold on blue clay - a false 'bottom.'Thompson and Noble struck gold on weathered schist - the true 'bottom:'

Pacey and Nieper were forced to turn back to the Manor Burn for water, and they brought in a low-level race from this stream to a point on the river flat below their claim. They then constructed a dray road over half a mile long for transporting their wash down to the race where they intended to sluice it. Within a month of having taken over the claim, Pacey and Nieper were box sluicing on the Manorburn flat and their returns were very good — 13 oz from 16 loads, and 31 oz from 24 loads from a 'paddock' 36 by 10 feet (11 by 6 m) which was excavated in three day's work.

A reporter who inspected the diggings on 16 August, was so impressed with what he saw that he referred to the place as 'a Coolgardie on a small scale'[4] —- an epithet which stuck.

By February 1895, Pacey and Nieper were still excavating by 'paddocking,' but as the 'bottom' of blue clay sloped down to the north-west, the depth of the excavation increased. Eventually the headwall was more than 16 feet (5 m) high, so they began tunnelling into the wash. Luckily, timbering was not required, as the overlying quartz grits were compact and stable. More stable, in fact, than the partnership between Pacey and Nieper.

It had perhaps been a strange alliance from the beginning — the miner Pacey and the butcher James Nieper. It was obvious that Nieper could not run a butchery business and a mine at the same time, so it was arranged that John Pacey, presumably a relative of William, would work at the claim in his place. In January 1895, the partnership broke up when James Nieper announced he could not continue to take an interest in two businesses. So he sold his half share of the claim, the water race with its four heads from the Manor Burn, and two dams, to Harry McDonald, one of the original prospectors, who had returned to the district. The price was £40, although it had cost Nieper £100 when he bought into the partnership just six months before. But then he had done well out of the claim.

John Pacey finished up also, and was paid £42 in wages owing by William Pacey, who then tried to recover half this sum from James Nieper. The question apparently was whether the wages were to be paid by the partnership or by James Nieper. The matter could not be resolved in the Magistrate's Court as the magistrate decided that his court did not have jurisdiction — as a partnership matter it should have been dealt with by the Warden's Court. He struck the case out, with costs of £3 against James Nieper. When the Warden's Court finally heard the matter several months later, Pacey got his money from Nieper.[5]

After a few months working with McDonald, Pacey decided to get out also. He bought the Criterion Hotel in Alexandra and Henry Fuller, Lewis Cameron and Martin Paget acquired the claim. The new owners had plans for bringing water to the claim for sluicing. Without water for sluicing, mining on the Manorburn diggings was hard work. The tunnelling in the hard compacted wash, the loading of drays and carting down to the race where it was shovelled through the sluice boxes, was time consuming, and put a cap on the amount of gold that could be won. But still no water was brought on to the claim.

Rapid changes in the members of a partnership usually meant that a claim was coming to the end of its life. So it was with the Prospectors' claim. By early 1896 the partnership had changed to George Todd, A McLean and Conway, and by the end of 1896, newspapers were reporting that the Manorburn field could be considered worked out unless someone with capital was able to bring in an adequate water supply to the upper part of the terraces to allow large scale hydraulic sluicing.[6] That someone was James Rivers.

James Rivers and 'Richmond Hill.'

James Rivers, a well-known storekeeper, local body politician and mining investor had acquired the Tucker Hill claim of William Jackson and George Campbell at the mortgagee sale. With the claim came a 18 mile (30 km) water race from Speargrass Creek, a tributary of Little Valley Creek which is, in turn, a tributary of the Manor Burn. Rivers built a substantial dam in Speargrass Creek which held back an extensive reservoir, but to ensure an adequate supply, he constructed a 6 mile (10 km) race from Mt Campbell Creek, another tributary of Little Valley Creek, and fed the water into his main race. This race brought water to the crest of the ridge overlooking Alexandra, and here some water was diverted into a large concrete tank from where it was led into a pipeline which supplied Alexandra with domestic water.[7] The bulk of the water fell down the front slope of the ridge in a series of waterfalls, before being picked up by a lower level race and conveyed to the Tucker Hill diggings. With the help of his manager, George Campbell, Rivers mined the Tucker Hill claim for several years. But eventually, as the ground was worked out, the gold output declined. Rivers looked around for another claim.

Rivers, in 1904, applied for a 5-acre claim at White Hill, together with a tail race to the Manuherikia River via the lower part of Edwards Gully. In granting the application the warden pointed out that if at any time the discharge from the tail race caused damage to the nursery of Howden and Moncrieff, or to the operations of the Manorburn Gold Dredging Company, then Rivers would be liable to pay compensation, and the Warden might impose further conditions. James Rivers called his claim 'Richmond Hill' after his birthplace in Middlesex in Britain.

At last someone was able to bring water on to the claim in sufficient quantities for hydraulic sluicing. Rivers simply diverted his Tucker Hill water into a race that led to a point overlooking the claim, and there it was fed into a line of pipes which, as was the practice, were supported on

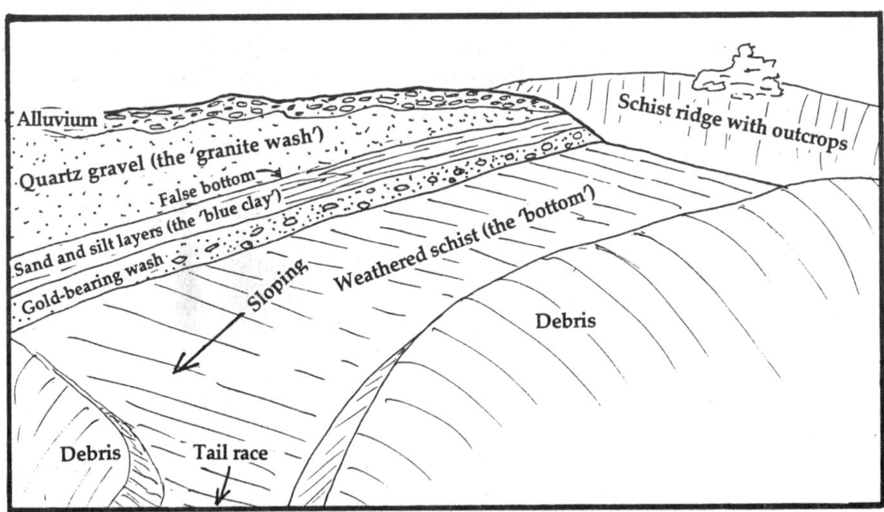

Figure 15.5. James Rivers' 'Richmond Hill' claim at White Hill.
Upper. The claim in the early 1950s before quarrying of sand had started.
Lower. A diagram, loosely based on the photograph, shows the various layers mentioned in the description.

low pylons of schist slabs.

There was sufficient pressure for effective sluicing[8] but almost immediately complaints were laid by the nursery, the dredging company and other miners that Rivers' tailings were causing nuisance, and that he was not complying with the terms of his licence. Rivers promised to be good. But within a few months he was again up before the warden facing the same complaints. It was ruled that tailings from his tailrace must go into a settling basin instead of directly into the river.

The water supply from Speargrass Dam was unreliable because the long race built through steep, rocky terrain was subject to much leakage, so, in 1905, Rivers decided to build a dam in the upper part of Edwards Gully. This would fill overnight and provide a constant supply during the day. The dam, a stone structure 60 feet (18 m) long and 20 feet (6 m)

Figure 15. 6. Looking down Edwards Gully with the dam built by Rivers at lower left. The race leading from this dam to the 'Richmond Hill claim is on the hillside to the right. The site of an earlier dam built by Campbell and Jackson, is marked by the poplar tree in the centre of the photograph.

high, formed a reservoir about 100 feet (30 m) long. It was situated about 200 yards further up the gully from an earlier dam built by Jackson and Campbell to store water for their Tucker Hill mine. This early dam was now providing a storage reservoir for Alexandra's water supply.

When Rivers applied to have his tailrace licence renewed in early 1906 there were a number of objections, and although the warden granted the licence, he made more conditions. Rivers must not allow his tailings to

Figure 15.7. The sand quarry in White Hill from the air. At the far right of the excavations is the original 'Richmond Hill' claim of James Rivers.

block the Manor Burn and so cause water to back up and flood Howden and Moncrieffs nursery. If necessary, he was to open up a new channel for the Manor Burn.[9]

The depth of overburden continued to increase steadily as sluicing followed down the sloping schist floor, so it was decided to extract the gold-bearing wash by means of a number of drives into the face of the excavation. Eventually even this became too difficult and the claim was closed down in January 1909.

The mine lay abandoned for forty years until tradesmen in Alexandra discovered that the yellow quartz sand was ideal for concrete work, and especially for plastering. Although the 'White Hill' has been partly destroyed by the growing demand for aggregate, it is still possible to recognise a number of features associated with it. The two stone dams in Edwards Gully, and the water race leading from the upper dam, are still to be seen. The place where the water was led into the pipeline, and a few of the stone pylons that supported the pipes can be recognised, as can the site of the original workings which gave rise to the rush to the 'Central Otago Coolgardie.'

NOTES

1. *Dunstan Times* 20 July 1894. Miners Rights would not be obtainable on a Sunday, presumably they were obtained next day.

2. The name of the stream is written as 'Manor Burn' but when the name is used as an adjective, as in 'Manorburn terraces,' it is usually written as one word.
3. Presumably named after John Edwards, who spent his mining life working on the Manorburn Terraces. He pegged out a claim on White Hill during the rush. Eventually he was found dead in his hut that was on a 'Residence Area' near the junction of the Manor Burn with the Manuherikia River.
4. *Dunstan Times* 24 August 1894. A reference to Coolgardie in Western Australia which was in the news at the time as a producer of fabulous quantities of gold.
5. *Dunstan Times* 5 July 1895.
6. *Dunstan Times* 6 November 1896.
7. For a description of this water supply scheme and the relationship between the Alexandra Borough Council and James Rivers see Chapter 16 in McCraw, 2000.
8. For a full discussion on James Rivers' mining activities around the Manor Burn see McCraw 2000 Chapter 22.
9. Details of Rivers' various applications and the conditions imposed are taken from the records of the Alexandra Wardens Court held in the Dunedin Regional office of National Archives.

16.

BONANZA!

— Gold to Agriculture

—

For a few years after the Dunstan Rush construction of large water races was undertaken at a frenetic rate. Dunedin investors in particular, were busy floating companies to bring water from distant streams, not with the intention of themselves using the water to mine, but of selling it to miners.

Ida Valley Water Race Company

In October 1865, a group calling itself the Ida Valley Water Race Company announced that it had sold most of the 40 shares (at £50 each) it was offering, and was preparing to construct its race. Its plan was to bring water from Moa Creek, at the south end of Ida Valley, to Lows Saddle on the Raggedy Range between Galloway and Ida Valley. Here the water would be divided into two races. One would run along the Galloway (the north-western) side of the range to the diggings at Blacks (now Ophir), and the other would be cut along the Ida Valley (south-eastern) side to the recently discovered Blacks No 2 and No 3 diggings.[1] The company was in no doubt that its water would be in great demand.

Fortunately, before much work had been done, it was discovered that the summer flow of Moa Creek shrank to as little as two heads, so the plans were changed. It was decided to use the upper Manor Burn[2] as the water source. This would require a race about 20 miles (33 km) long, including some natural gullies, and it would carry 20 heads of water. Such a large undertaking would require more capital, so on 1 March 1866 a new company was registered with a capital of £8,000 in £25 shares. About half of the 29 shareholders were local to Ida Valley, but at least eight came from Dunedin. Those with shares in the old group were invited to transfer to the new company.

Construction work on the race, which the newspaper described as 'the best piece of work in the district,' was pushed ahead under the management of the experienced miner, Andrew Wood. By September 1866

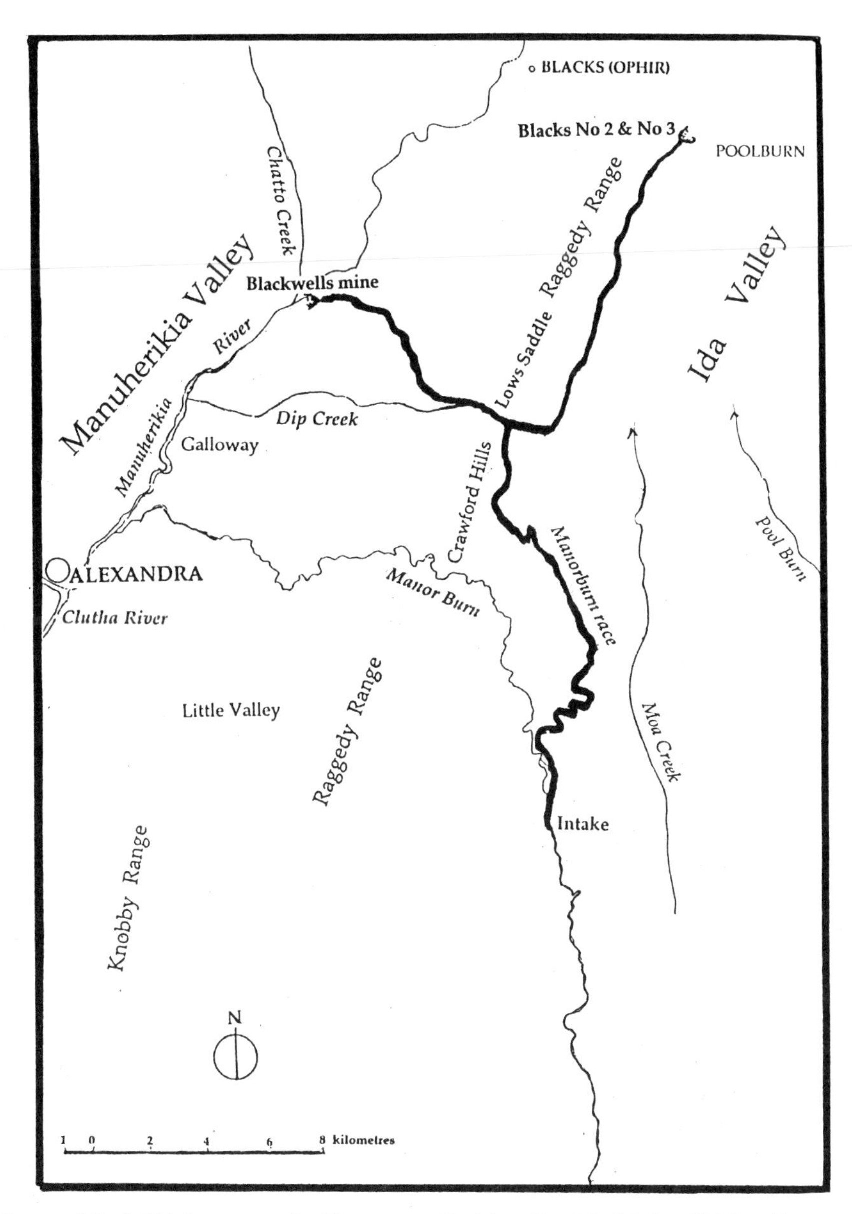

Figure 16. 1. Water races built, or supplied by, the Ida Valley Water Race Company and its successor, the Manuherikia Water Race Company.

he was able to report that over 16 miles (26 km) of race had been completed, but there were still about three miles to construct to reach Lows Saddle, an intake to build on the Manor Burn, and some fluming to erect.[3]

Miners at Blacks, in particular, were looking forward to the arrival of the race, as shortage of water was a great hindrance to development of mining at this locality. So they were more than a little alarmed when the company apparently changed its plans and decided to do some mining itself. Before the race to Blacks was constructed, it announced, a race would be built to take water on to a high terrace, a short distance east of Galloway Station homestead,[4] where the company had acquired six acres of rich ground.

This change of plan was actually a desperate move to make some money quickly and defuse the mounting dissatisfaction among shareholders. Many had bought shares with the sole object of making money with the expected rise in share prices. Unfortunately, this didn't happen—the £25 shares dropped to £5. In the unstable market conditions, investors in the company lost confidence and refused to meet their calls. In January 1867 a number of prominent shareholders were successfully sued by the company for the money owed.[5] As can be imagined, this caused something of a fuss and a meeting of angry shareholders was held which demanded, and received, the resignation of the directors, manager and secretary of the company.

It was all too late. Already assets were having to be sold off to meet pressing debts, and in June 1867 the company went into liquidation. Newspaper editorials of the time bemoaned the fact that the expenditure of only a few hundred pounds more would have seen the race completed, and not only begin to generate income but also provide much needed work for many miners.[6]

Manuherikia Water Race Company

At the inevitable auction sale, the race and water rights, on which £7,000 had been spent, were bought for £450 on behalf of a few of the Dunedin shareholders of the defunct company. They had formed a new company, with the hope of salvaging something from the wreckage of the old. The Manuherikia Water Race Company was registered in November 1867 with a capital of £3,500 and seven shareholders. Tenders were immediately called to finish the three miles of race to Lows Saddle. When this work was completed the company could divert water into natural watercourses on either side of the saddle. The one on the Galloway side, 'Wet Gully,' is now known as Dip Creek, but that on the Ida Valley side is unnamed. It was advertised that water was available for rent, but would be free to miners prospecting on either side of the saddle.

One of the first to rent water was William Blackwell, who in late 1868, had taken up a patch of rich ground high on the terrace above the Manuherikia River, opposite the mouth of Chatto Creek. He cut a race about six miles (10 km) long to his claim from a point in Dip Creek about a mile below Lows Saddle. He paid £12 per week for three heads of water, a price regarded as high by the locals, but he was on good gold, returning £15 to £20 per man each week. Low, of Galloway Station, also rented

Figure 16. 2. Blackwell's Diggings.
Upper: Remains of the race William Blackwell cut in 1868 to convey water from Dip Creek to his claim opposite the mouth of Chatto Creek.
Lower. Blackwell spent several years ground-sluicing a patch of terrace gravels high above the Manuherikia River.

water for stock water and irrigation. Within a short time the promised race to supply Blacks No 2 and 3 diggings had been cut along the eastern foothills of the Raggedy Range.

Rockside Mine—A Family Enterprise

William Hansen, a Norwegian, had worked around Alexandra for 10 years, mainly as a butcher, but he was also interested in mining. In 1880, with his brothers-in-law George and Lewis Cameron, and later Henry Symons, another brother-in-law, he opened up a claim on Dumbarton Terrace opposite the new Galloway Station homestead. It was close to the spot where the old Ida Valley Water Race Company had intended to mine. The partners bought the Manor Burn race from the Manuherikia Water Race Company, which, with the decline in mining, was only too willing to

Figure 16. 3. Partners in the Rockside mine.
Clockwise from Top: William Hansen, George Cameron, Henry Symons, Lewis Cameron.

dispose of it, and cut a race from Dip Creek down to their claim. Although the main race from the Manor Burn had been built to carry 20 heads of water, leakage and lack of maintenance had reduced its carrying capacity over the years until only about five heads were delivered into the dam near their claim.

The claim, of two acres, was on the edge of a gravel terrace that stands about 70 metres above the flood plain of the Manuherikia River, and stretches upriver from Dip Creek for more than three kilometres. It is part of a once much larger area of high-level terrace which occupied most of Galloway Flat, but this remnant has been protected from river erosion by a ridge of rock which lies between it and the river.[7] The gravel contained a little gold throughout its thickness but was richer near the base. Hansen admitted that the partners recovered gold valued at over £7,000 over seven years.

Shortage of capital held back development. The working face was between 40 and 50 feet (12 to 15 m) high, so a very large amount of gravel had to be removed to reach the richer wash beneath. Disposal of the huge amounts of tailings was a problem that led to disputes with the County over obstructing the road leading to Ida Valley and Dunedin. However, the main difficulty was, as always, the shortage of water. The partners decided to build a dam in the headwaters of the Manor Burn that would convert the extensive Greenland Swamp into a large reservoir. If this were done, the race from the Manor Burn would need to be enlarged to carry a greatly increased flow of water. To give more weight to their applications for these mining privileges, they formed themselves into a private company.

Galloway Hydraulic Sluicing Company

The Galloway Hydraulic Sluicing Company was formed in early 1889 with shareholders William Hansen, Lewis Cameron, George Cameron and Henry Symons. Its first application was for a 50-acre (20 h) block running back on to the terrace from the original claim. Then the company applied for, and was granted, a licence for a dam on the Manor Burn, a reservoir site of 1,500 acres covering the Greenland swamp, and the right to 50 heads of water from the dam. Unfortunately, forming the company did not produce the capital necessary to carry out these works.

Disappointed, Hansen began to lose interest in the whole enterprise. Besides, he was developing interests in other mining concerns, including the acquisition of a large block of adjacent land suitable for dredging. So in 1893 the shares of Hansen and the Cameron brothers were put up for auction sale but failed to sell. Then George Cameron died and his share was taken over by his mother, the redoubtable but highly respected Sarah Cameron. Lewis Cameron, like Hansen, went off to other fields and 1898 saw Henry Symons left to operate the mine, although the others still held their interests in it.

Figure 16. 4. Sluicing in progress at the Rockside mine, Galloway, during the 1880s. The sluicing nozzle is helping to force gold-bearing wash through the sluice boxes on the left.

their interests in it.

The claim and water races remained on the market, and finally attracted a group of Dunedin businessmen who began negotiations to purchase. One condition was that the titles to the water rights, races and claims should be clear of any encumbrances. With this assurance given, the group formed a public company, the Bonanza Gold-dredging and Sluicing Company Ltd, which was registered in September 1899.

The Bonanza Gold-dredging and Sluicing Company Ltd.

The transfer of the assets from Hansen and party to the Bonanza Company was not without incident. Just before the deal was completed, a claim for 'Forfeiture' was lodged against The Galloway Sluicing Company by some unknown party working through an agent, Mr Mackney. The grounds were that the licences for the water races and water rights had been allowed to expire and had not been renewed within the prescribed 60 days.[7] It was something of a technicality but could have delayed or even scuppered the whole transaction. At the same time Mackney, on behalf of his principals, applied for the dam site, races and water rights, but consideration of this application was adjourned until the Forfeiture question had been settled. The Warden reserved his decision.[9]

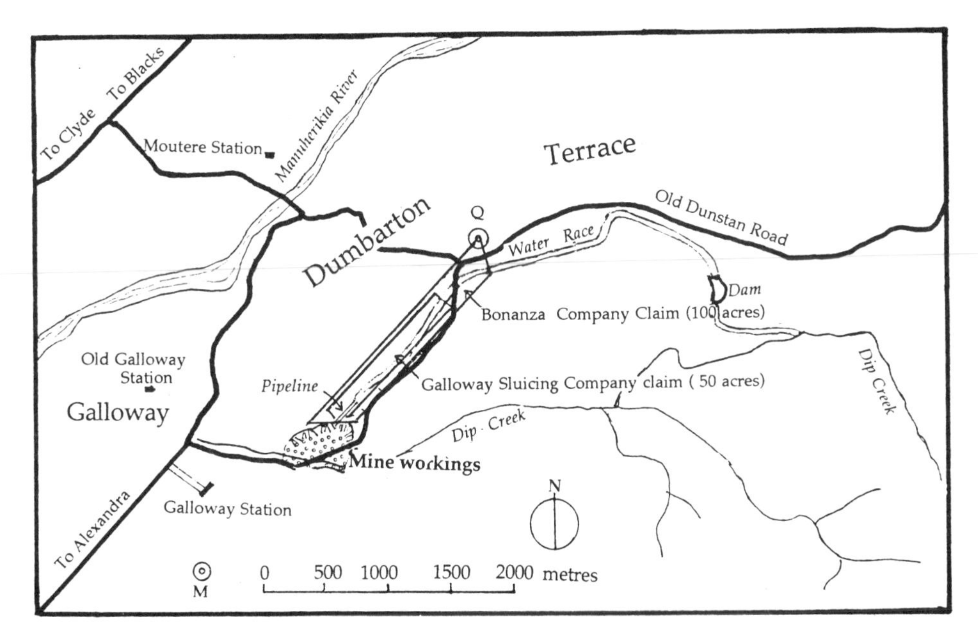

Figure 16. 5. Locality of Rockside mine with later mining claims and other main features.

Ten days later at the first meeting of shareholders of the new company, the chairman, G. Fenwick, made what he described as 'a few observations' about this attempt to 'jump' the vendors' water rights and mining privileges.

> . . he was sorry to say the attack had come from those from whom it should not have been expected on account of their position in the community. A jumper was not usually a person who held anything of a position in the community, yet in this instance there were 17 or 18 business and professional men in Dunedin who had banded themselves together, and employed an agent, who had made efforts through the Warden's Court to harass the owners of the claim, and to get possession of their dam site and water rights.[10]

The *Dunstan Times* was much more explicit. and asked about the credentials of Mr Gordon Mackney, the plaintiff in the case:

> An interest was added to the case by the question being raised as to the existence of such a person as Mr Mackney, the plaintiff, or, granting his existence, whether he were a mere dummy for Mr Rawlins [Member of the House of Representatives] and not *a bona fide* interested in the matter.[11]

This adverse publicity must have had an effect, for when Mr Mackney was called to speak to his application to take over the mining privileges on the Manor Burn, he did not appear, so his application was struck out.

The Court not only found in favour of the Galloway party but also issued them brand new titles, under the latest 1898 Mining Act, for the 20-head water race and the 1,500-acre dam site. This enabled the water rights, races, dam site and claim to be transferred to the Bonanza Company, with a clear title, for cash and shares to the value of £3,000. The Warden also approved the Bonanza Company's application for a further 50 heads of water from the proposed Greenland Swamp dam.

The Company seemed surprised that its application for the extra 50 heads was successful and was caught somewhat off guard. But it quickly set out its plans. It was decided to build the Greenland Swamp dam immediately, and to enlarge the main race from the dam to carry the extra water. This involved increasing the capital of the company to £30,000. The 50 acre claim, taken over from the Galloway Hydraulic Sluicing Company,

Figure 16.6. The dam that converted the Greenland swamp into a reservoir was built in 1901-1902 by the Bonanza Company. It was submerged when the Upper Manorburn Dam was built but reappears when the water level is low. The maximum water level is clearly seen on the bank in the background.

was enlarged to 100 acres. A programme of test boring was commenced at the claim, followed by plans for extensive sluicing, which would remove much of the rather barren overburden. Then a relatively small dredge would be built which would deal with the exposed wash.

The race work and dam building were started with 40 men in March

1901 and completed in 1902, but when sluicing commenced it was soon evident that gold returns were too low to sustain a company with such a large capital investment. Many excuses were presented at shareholders meetings — the sluicing equipment was too small to move the great quantity of overburden, the sloping bedrock meant extra large and expensive tail races had to be built, and dissatisfaction with management meant three managers (including William Hansen) in as many years. But the fact remained that the claim was unprofitable.

In April 1903 the company decided to abandon mining, including any thoughts of building a dredge and, with a full turn of the wheel, concentrate on selling its water to other miners just as the original Ida Valley Water Race Company had planned 35 years before. Advertisements called for those who wanted water, to contact the company.

It was about this time, mid-1903, that a petition was drawn up by the people of the district, asking the Government to buy the water rights and races of the Bonanza Company and use the water for mining and irrigation, particularly on the extensive terraces east of Galloway Flat. Government replied that it would be better if the water rights were taken over by local interests.

It was quickly realised by the company that selling water was not as simple as it had appeared. To supply local miners with water would mean constructing a network of water races, and the company was not in a financial position to do this. But one group who held a number of claims along the north bank of the Manor Burn, and who would later form the Manorburn Sluicing Company, put up a particularly strong case for water. As a result the Bonanza Company applied to build a race, over six miles (8 km) in length, along the foot of the Crawford Hills. To pay for this work 3,000 preference shares were offered, but there was little enthusiasm amongst investors.

Meanwhile a new manager had demonstrated that by concentrating all the available water on a small part of the mine, he could recover sufficient gold to cover costs with a little to spare. The directors were enthusiastic about the improved returns. So when, as a result of the petitions, the Government asked in February 1905 whether the company wanted to sell, the Directors replied that they were not interested in such a proposal. This led to a little embarrassment when a deputation of local people met Mr Seddon, the Prime Minister, at Ophir in March and appealed to him to buy the Company. Seddon produced the letter from the company and pointed out that the Government could not force an unwilling party to sell.

There is no doubt, however, that the figures produced at the Annual Meeting of the company, which showed £19,000 had been spent to recover £1,167 worth of gold since registration, changed the shareholders' minds. They instructed the directors to offer the assets to the Government for £20,000.

Figure 16. 7. The Bonanza mine as it appears today. The mine has been used as a gravel pit for some time so parts of the mine face have been considerably modified.

A Messy Deal

This offer brought the Chief Inspector of Mines down from Wellington. He made a quick examination and went back to report that, in his opinion, the assets of the company were worth only £12,000. The company responded by asking for £18,000. The Government then asked for a detailed list of the Company's property and when this was supplied in July, Mr Gordon, the Chief Inspector, came back and spent a week in September examining everything very carefully. As a result of his report the Under-secretary of Mines wrote to the company on 1 November 1905:[12]

> I am directed by the Minister of Mines to offer, without prejudice, the sum of £13,050 for the undermentioned privileges, namely dam at Greenland Swamp, main race, flume, bridges, dam at head of Wet Gully, right of Blackwell's race, right of the Ida Valley race, and tailraces and on execution of transfers of these privileges to His Majesty the King, to the satisfaction of the Crown Solicitor at Dunedin, the sum offered will be paid.

After the directors had consulted the shareholders, the company's unconditional acceptance of this offer was telegraphed on 3 November and confirmed in writing on 6 November. Then things started to go wrong. Only a week later a letter came from the Under-secretary of Mines:

> In reply to your letter of the 3rd inst accepting the offer. . . I am directed to inform you that further consideration of the matter is held over for the present.

The letter, which arrived from the Under-secretary at the end of the month, left no doubt:

> Referring to my letter of the 14 inst I am directed to inform you that the offer to purchase the mining privileges . . .is withdrawn, and further negotiations for the purchase of the properties are now at an end.

The company wrote to the Under-secretary asking for an explanation as to why the Government wanted to break what the company regarded as a contract. At the same time, the directors sought an interview with the Premier who had arrived in Dunedin on 30 November. Mr Seddon was reluctant to meet them but when he finally did, he admitted he did not know why the offer had been withdrawn. He could only suggest that officials had noted the rise in price of the company's shares after the Government offer was made, and had interpreted this to mean that the offer was too high. He also pointed out that the company could not bind the Crown, even if they were thinking of it. Finally he did say that he chose to regard the situation as 'negotiations suspended' rather than abandoned. When the Premier returned to Wellington he wrote confirming his view that negotiations were suspended, and also that he was putting the whole matter in the hands of the Crown Solicitor in Dunedin who would deal with it.

The Company's reply to this was to point out that as far it was concerned the matter was not one for 'negotiations.' There was a contract, and if the Government failed to honour this then the Company would place the whole matter before its shareholders — in other words, 'go public.'

A short time later, the directors of the company met Sir Joseph Ward who admitted that the withdrawal of the offer was a deliberate Cabinet decision, and was based on a communication received from Dunedin. It contained certain allegations that had to be inquired into. He would not disclose their nature because of confidentiality but did say that

> . . . if they were true he did not care if the matter of the Government's action in respect of the contract were published to the whole world.

Knowing that the Annual General Meeting was coming up in March the Secretary had asked Government, some time previously, for a definite answer as to when the settlement would be made. Government replied that two engineers would be sent down to prepare a further report on the property. By the time of the meeting this report had been in Government's hands for a fortnight, and still there was no word.

On the morning of the AGM, as a result of a telegram to the Minister of Mines, the Company received an assurance that a definite offer, which would include the claim and mining plant, would be in the Company's hands by that evening.

At the Annual General Meeting of the Alexandra Bonanza Gold

Figure 16. 8. The Alexandra race running along the foot of the Crawford Hills. In the end it supplied very little water to miners - most was used for irrigation, as seen in the photograph.

Dredging and Sluicing Company held on 1 March 1906, questions were asked, as expected, about the sale. One shareholder wanted a definite assurance about the sale. 'All I want to know is are we going to sell our water privately or have the Government bought it?'

The Chairman replied that they had no settlement at the moment but the Government had undoubtedly bought the water rights. He then read out the letter of 1 November in which the Government made the offer, which the company had accepted. Then he told them about the subsequent withdrawal, the pressure put on Government to honour the contract and the offer which was to arrive that evening.[12]

It seems that the offer did arrive that evening. It was for the same sum of £13,050 but this time included the claim and the mining equipment, so really was a lower offer than the previous one, which was for the water privileges only. At a meeting called the following week the shareholders accepted the offer. But another three weeks were to pass before a telegram from the Minister of Mines informed the Company that Mr Hayes, the Chief Inspector of Mines, was in Dunedin to effect a settlement of the sale.

As a final gesture, the Company decided to print the whole story of the negotiations between the Company and the Government, including the

Figure 16. 9. The Upper Manorburn Dam

Upper. Construction begins. The Manor Burn has been diverted into a flood race (left) by a temporary dam. The pipe down from the right presumably delivers water under pressure which would be used either to operate a pump to lift water, or perhaps gravel, from the excavation for the dam foundations.

Lower. The completed dam is 29 metres high and 118 metres along the crest.

text of the letters, and distribute it to shareholders and the press. The *Otago Witness* headlined the article, 'The Sacredness of Contract'.[13]

At long last the Government had acquired the valuable assets of the Bonanza Company, which was left with little to do except go into voluntary liquidation. It did so in early April 1906.

Figure 16. 10. A considerable amount of work was undertaken by the Government enlarging the old Bonanza race and lining leaking sections with concrete. The upgraded race could carry about 100 cusecs (3 cubic metres/ second).

His Majesty the King— Race Owner

It is doubtful whether anyone informed King Edward VII that he was now the owner of several water races and sundry other water and mining privileges in a remote corner of New Zealand. As far as we know he never attended the Warden's Court in Alexandra in support of his applications for the Privileges. Nevertheless, he was the licensee named on the new documents prepared by the Court. The King was fortunate in having officials of the Mines Department, and later of the Public Works Department, to administer the water races for him.

Even before the transfer was complete, farmers and miners were applying for the water, and almost immediately the survey of the race along the foot of the Crawford Hills was put in hand. This race, which it was decided would be called the 'Alexandra' race, was six miles (8 km) long and eight feet (2.4 m) wide, 18 inches (45 cm) deep and cost about £2,000. About 50 men were employed and the work was completed in May 1907.

The main customer for water for mining on the Galloway side of the range was expected to be the Manorburn Sluicing Co., made up of an amalgamation of claims held by a group of investors. But this company faced an injunction taken out against it by Howden and Moncrieff, who feared that tailings from its activities would flood their nursery at the mouth of the Manorburn. In spite of a campaign by the local Miners Association to have the Manor Burn declared a 'Sludge Channel' into which miners could freely discharge tailings, Government decided, instead, to build a weir to trap the tailings. The delays caused by all this led to a successful 'attack' on the claims of the Manorburn Sluicing Company by James Rivers, on the grounds that they had been abandoned. The result was that the claims were declared 'Forfeited.' In the end only three miners signed up for water from the Alexandra race.

Ida Valley Irrigation

Over in the Ida Valley, on the other hand, large numbers of farmers were willing to sign up for irrigation water, and they were supplied as far as possible from the water stored in the Greenland dam. But there wasn't enough to satisfy demand, and agitation began to have the old 20 foot-high dam at Greenland Swamp increased in height to allow more storage. The matter was raised in the House of Representatives in 1910 by Mr Robert Scott, the local Member, who pointed out that the present dam held less than three month's storage, and unless this was increased the Government had wasted its money in buying the expensive Bonanza assets. Government had to make up its mind whether it was serious about developing irrigation.[14]

Government had, however, made up its mind. It had already earmarked the sum of £100,000 in 1906 for development of irrigation, and had allocated £25,000 for Ida Valley.

Before much further progress could be made, it was necessary to find out how much land was suitable for irrigation and where it was situated. In 1909 a comprehensive survey of Central Otago was carried out by Messrs J. L. Bruce, an officer of the Department of Agriculture, and J. H. Dobson, an engineer.[15] Included in their report was a list of 60 farmers holding some 32,954 acres (13,324 h) who were willing to take water at 10 shillings a week per head of water.

As a result of Bruce and Dobson's report, it was decided that the southern part of Ida Valley would be the first to be irrigated. But before large sums of money could be spent, the Public Works Amendment Act of 1910 required Government to assure itself that any scheme would be self supporting. This meant almost all the farmers had to enter into long-term contracts to take and pay for the water. When this had been achieved, the search for suitable dam sites commenced. There never was any intention of raising the old Greenland Swamp dam, and it was not long before sites were selected for big new dams in the Manor Burn and Pool Burn.

Figure 16.11. The Manorburn race.
Upper. The intake in the Manor Burn about four kilometres below the Upper Manorburn dam.
Lower. An air view of the Manorburn race as it is today.

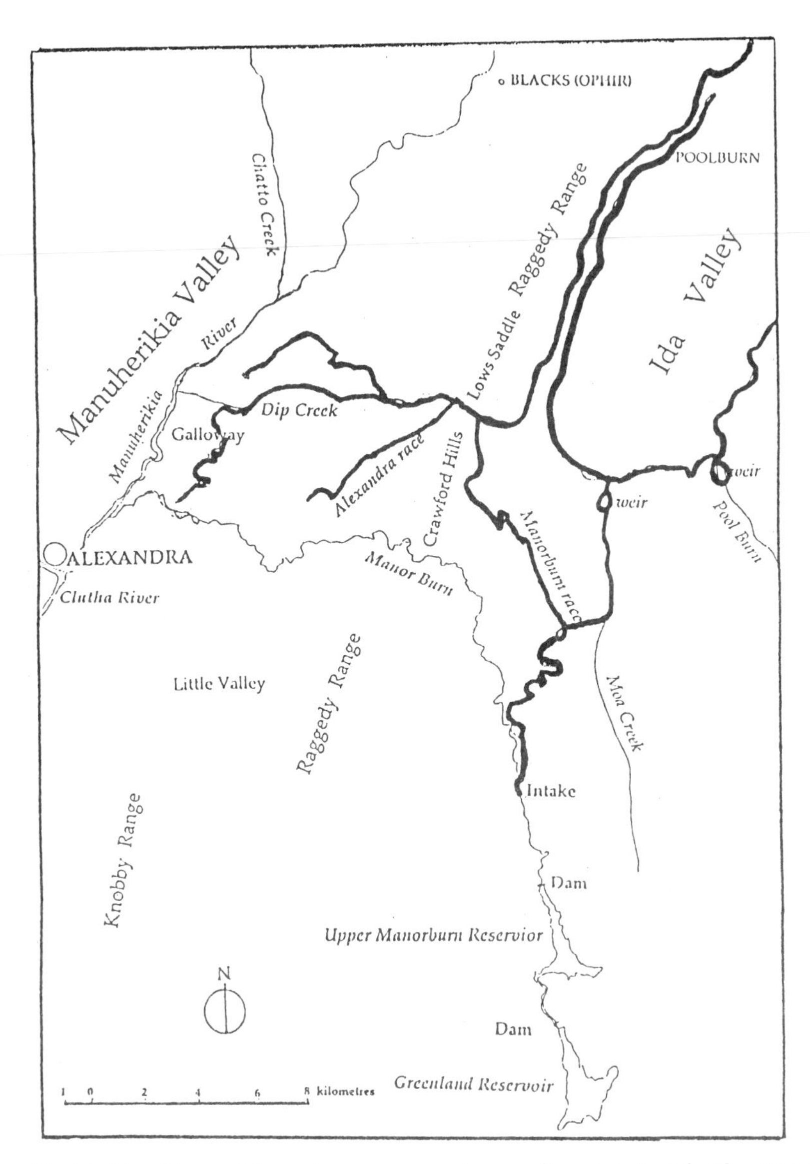

Figure 16. 12. The Upper Manorburn reservoir supplies a network of water races Irrigating the southern part of Ida Valley and Galloway in the Manuherikia Valley.

The old Bonanza race from the Manor Burn was renovated and enlarged, the race to Blacks No 3 was enlarged and extended, and a new race cut around the eastern side of Ida Valley which picked up additional water from Moa Creek. Irrigation was extended on the Galloway side to include much of Galloway Flat.

At the same time plans were being prepared for a very large dam and reservoir in the upper Manor Burn. Construction commenced in 1913 and was completed in 1915, in spite of difficulties of working with concrete in very cold temperatures, and with shortages of materials during the Great War. The old Greenland dam was submerged by the huge new reservoir which covered 706 hectares.

So the original 1865 plan of the Ida Valley Water Race Company was fulfilled, in that water from the Manor Burn was eventually delivered to Galloway and Ida Valley, not to miners as the original company had planned, but to farmers. And it is ironical that it was the failure of a later company which had chosen the evocative name 'Bonanza,' that led to a bonanza, not in gold but in farm production.

NOTES

1. Blacks No. 1 diggings were near Blacks (now Ophir) township and Blacks No. 2 and No. 3 were on the other side of the Raggedy Range on the slope overlooking the present settlement of Poolburn.
2. The official name of the stream is 'Manor Burn' but it has become the practice to use 'Manorburn' for the names of the locality, dams and water races, etc. eg. Manorburn Flat, Upper Manorburn Dam, etc.
3. *Dunstan Times* 9 September 1866.
4. Galloway homestead was at this time about a kilometre northwest of the present homestead.
5. *Otago Daily Times* 22 January 1867.
6. *Dunstan Times* 8 June 1867.
7. J. Park, who published a report on Hansen's claim in 1890, thought that the gravel filled an old river course. This is most unlikely because the gravel rests (according to Park) on Tertiary sediments and, in addition, there is present the layer of bright red clay which is believed to represent an ancient soil that developed on a thick layer of soft, deeply weathered schist. None of these materials would survive river erosion. It is more likely that the structure is a small down-faulted depression.
8. This was widely regarded locally as a case of 'claim jumping,' that is, taking possession of someone else's claim. In the early days this was often achieved by physical force but later sharp-eyed lawyers using devious technicalities did it.
9. *Otago Witness* 14 September 1899.
10. ibid 21 September 1899.
11. *Dunstan Times* 15 September 1899.
12. *Otago Witness* 7 March 1906.
13. *Otago Witness* of 25 April 1906 sets out the history of the negotiations.
14. *Dunstan Times* 26 September 1910.
15. Bruce and Dobson 1909.

17.

TURNING POINT

—Nursery Defies the Dredges

George Howden received his early training as a seedsman and nurseryman in Scotland and came to New Zealand in 1878 to work for James Laird, the well-known nurseryman of Wanganui. Howden did not like Wanganui, and after a short time, moved down to Dunedin. Here he met up with John Moncrieff who had also been trained as a seedsman in Britain. Since arriving in Dunedin he had been working for Messrs Nimmo & Blair Ltd. The two young men joined forces, setting up as seedsmen and florists in the early 1890s.[1]

Realising that they could not properly cater for the needs of their customers without a nursery, Howden and Moncrieff bought out several small nurseries, including that of George Dick in North East Valley. Dick, who was not only a first class tradesman, but also understood nursery management and how to handle staff, became the nursery foreman.

The new firm prospered, and with fruitgrowing beginning to expand in Central Otago, the partners took the opportunity to establish an 'experimental' nursery at Galloway in 1895. Their purpose, they said, was to find trees and plants suitable for the region. Perhaps so, but there can be little doubt that the increasing population accompanying the growing number of gold dredges, and the rising interest in fruitgrowing as the railhead approached, influenced the decision.

The firm purchased the lease from James Rivers of a 34 acre (14 h) block of fertile but bare land between the Manor Burn and the Manuherikia River. They were just in time. By the following year the first dredge was at work nearby and every acre of the surrounding land had been pegged out for gold dredging. They would not have had a hope of leasing this block of land for horticulture had they delayed a year. On the maps, the more or less rectangular section that was the nursery land, stood out clearly from a sea of dredging claims.

The lease was for 21 years granted under the The Mining Districts Land

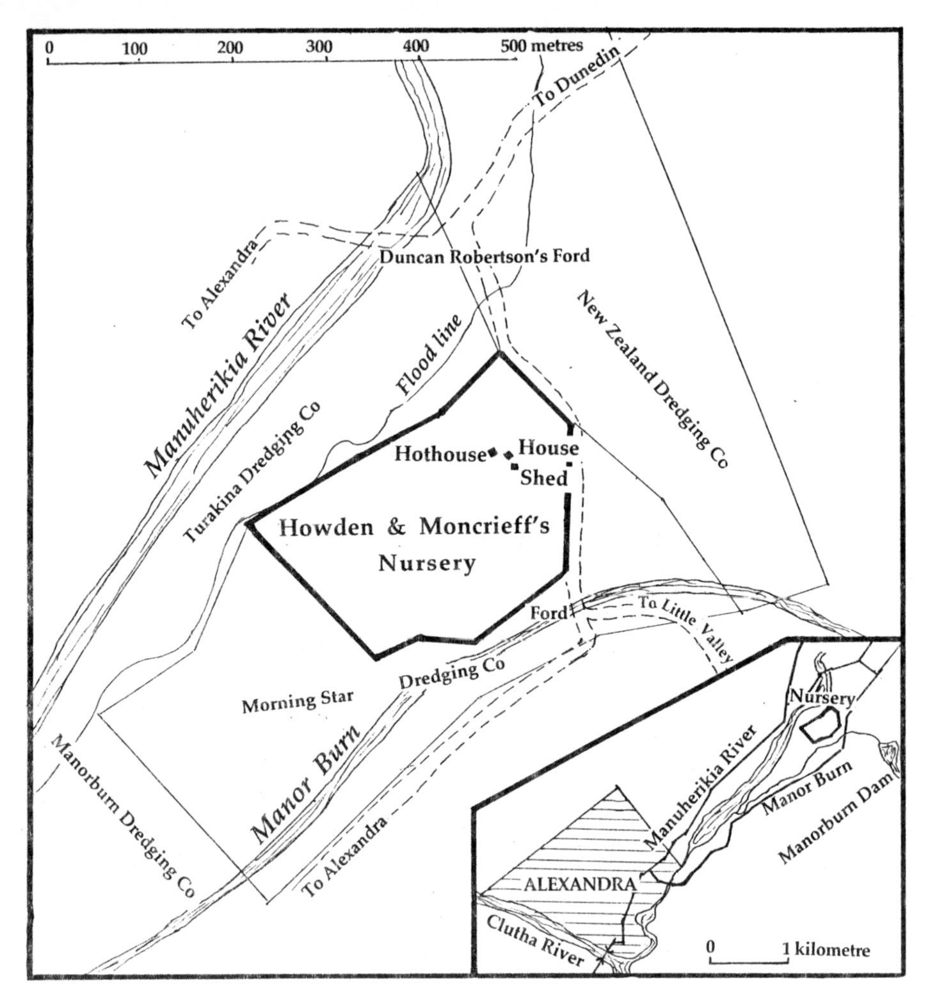

Figure 17.1. Location map shows Howden and Moncrieff's nursery surrounded by Dredging claims.

Occupation Act, 1894, and although the lease was granted by the Land Board, the application had first to be submitted to the warden. He had to determine whether the granting of the lease would adversely affect any mining interests. If the lease were granted, gold prospectors still had free access to the land and miners had the right, with the approval of the warden, to run water races, tramways and so on, through it, provided compensation was paid. But the most important provision of the Act was that applications could be made to mine the land, and if after a hearing, the warden decided to grant the application, the Crown could resume the land. The lessee would be entitled to compensation for 'substantial improvements.'

Howden and Moncrieff's nursery manager, George Dick, went up to establish and manage the new property. The speed with which the place was developed was subject of comment in the newspapers. Within a week of taking over the property, a water right of two heads from the Manor Burn for irrigation had been applied for and granted. Within a short time poplar trees for shelter were planted round the boundary, a house and hothouse were constructed and ground prepared for planting out the nursery.

By the end of the year the place had become an oasis on the dry, barren, Galloway Flat and a showplace for visitors to the district. A newspaper report[2] describes a large glasshouse 15 by 40 feet (4 by 12 m) holding 3,000 tomato plants and heated by hot water piped from a furnace. Outside were extensive plantings of rhubarb, strawberries and 2,500 vine cuttings had been laid out. In addition, a wide range of fruit trees, including mulberries, were available, as well as a large variety of ornamental trees.

Within twelve months of commencement, there were four acres devoted to the nursery which now contained 60 varieties of gladioli and some 400 'tea' roses. There was also a large plot of bulbs including tulips, narcissi, hyacinth and Spanish iris. Hundreds of fruit tree stocks were at the stage where they would be grafted next season and then offered for sale the following year.[3]

Later an extensive seed-producing operation (onion, parsnip, carrot and various flower seeds were mentioned) was developed which covered more

Figure 17. 2. Seeds drying at the nursery in 1903.

than 20 acres (8 h) of the property. The remainder was devoted to the nursery and to a permanent stone fruit orchard.

The Dredging Claim

In 1898, what Howden and Moncrieff feared most came to pass— an application to dredge their land was lodged with the Warden's Court. Immediately they sent in an objection. The applicant turned out to be Laurence Ryan, a Mining Agent of Alexandra, who was acting on behalf of a group of investors.

At the hearing,[4] the warden explained that he was required first, under Clause 9 of the Act, to decide whether the land was 'required' for mining. If it were, then he had to decide whether or not to grant the application. He pointed out that it could be taken for granted that the ground contained gold in payable quantities, and the Mining Act of 1891 made it clear that in a mining district, the mining industry was to be regarded as the paramount industry, but not the only industry to be considered.

It was universally conceded, the warden went on, that Central Otago was pre-eminently suitable for the growing of certain kinds of fruits, and it was all important that this industry be fostered in every reasonable way. The enterprise of the objectors was no doubt ultimately for their personal gain, but would materially assist the development of an industry that was yet in its infancy in the district. To grant the application might well result in the objectors giving up and leaving the district, thus strangling a growing yet struggling industry.

An inspection showed, the warden said, that the land occupied by the nursery was not blocking access to large areas of auriferous land beyond. Recent cases in the Supreme Court had clearly shown that it was not incumbent on a warden to grant every application that came before him. He could, on the grounds of public interest, refuse any application. On the grounds of public interest he then turned down this particular application.

Howden and Moncrieff were lucky. Very few applications to mine land had been turned down in this, a mining district. The case did show,

Figure 17. 3. Howden and Moncrieff's orchard at Galloway in 1913. Manor Burn flowing right to left in foreground.

however, that public opinion, as reflected in the warden's attitude, was undergoing a change. In spite of the frenetic gold dredging activity that was already gripping the district, sensible people realised that it could not be long-lasting and that industries such as sheep farming, and particularly fruitgrowing, offered more permanent and stable livelihoods.

Howden and Moncrieff's satisfaction in their victory must have been tempered by the severe dressing down they then received from the warden. At the end of the case he had this to say:

> I must however remark it is a thousand pities the objectors did not permit the Warden to arrive at this conclusion without invoking the aid of extraneous official pressure to which he has been subjected at their instance. That pressure has had no influence on the present decision . . . but the fact that it has been brought to bear, is to be regretted, because if Wardens are to be subject to influences of this sort their judicial independence will be at an end. . . even assuming this is a proper case in which to allow costs, I will make no order in that direction.

No clue is given as to who brought the pressure to bear or of what nature the pressure might have been, but the warden certainly made it clear that he was not pleased with what had happened.

Now that the warden had given his ruling, there was little point in other companies applying for mining rights over the nursery land, so George Dick, the manager, felt secure and got on with developing the property. In

Figure 17. 4. The sludge weir in 1928. The weir was built across the Manor Burn in 1908 to contain mine tailings and prevent them flooding over the nursery.

September 1899 he was able to donate 60 English trees to the Recreation Ground in Alexandra, thus laying the foundations of the public gardens.

Members of the Executive of the Otago Central Railway League were greatly impressed when they paid a visit in 1903, although they found the crossing of the Manuherikia River by 'chair'* and the two mile walk somewhat trying. By this time some 11 or 12 acres of the property were planted in orchard, and apart from the area devoted to the nursery, the remainder was given over to production of seeds. Seeds of parsnip, onion and carrot were drying on trays in the sun. The seeds were said to be of first rate quality, perhaps evidenced by the fact that the firm had orders for 10 cwt (over 500 kg) of parsnip seed when the total crop was only 400 kg. In another part of the garden, onions at the rate of 21 tons to the acre were being harvested while most of New Zealand's onions were being imported from California.[5]

Another Mining Threat

Shortly after the turn of the century, George Dick perceived the possibility of a new threat to the orchard. In 1903 several Dunedin businessmen applied for, and were later granted, extensive claims in the shallow gullies falling into the Manor Burn. Although the claims were adjacent, they were separate and held under individual's names. It was the intention, apparently, to work as a syndicate. The claims covered in total over 300 acres (120 h) of the low hills on the northern side of the Manor Burn adjacent to Galloway Flat.

It took a long time to organise the enterprise. First, water had to be obtained from the Bonanza Water Race Company. Midway through arrangements this company was taken over by the Government and negotiations had to begin again. By June 1908, when it was decided to form a public company, no mining had yet been carried out.

Howden and Moncrieff were concerned that the activities of the Manorburn Sluicing Company would cause the Manor Burn, already choked with tailings, to overflow and spill into their nursery at the point where the stream turned sharply to the south-west. So they and other nearby settlers opposed the granting of a mining license to the company.

The Government decided that, in the public interest, it would build a 'sludge weir' across the Manor Burn. This structure was designed to hold back the tailings and so protect the low-lying ground near the mouth of the stream, but at the same time allowing mining to go ahead. Having received this assurance from the Mines Department, and after inspecting the site, the warden granted the company a licence on condition that mining start within six months, that is, by 8 December 1908.

Howden and Moncrieff made it clear that if the company dared to start mining before the weir was built they would slap an injunction on it, so the company had no option but to wait until the weir was erected.

Figure 17. 5. A 1949 air photo shows no vestige of the nursery or orchard but the Willows and tall poplars around the boundary are still evident. Manorburn Dam at bottom of photograph.

The weir was built by the Public Works Department and was completed during February 1909. The Manorburn Sluicing Co. geared itself up for a start to its mining operations at the beginning of March. But it had overlooked something — the six months grace granted by the warden had expired and the company no longer had a licence to mine.

James Rivers, a well known merchant and mining investor of Alexandra, forthwith pegged out a 10 acre (4 h) claim in the middle of the Manorburn Sluicing Company's claim and it was awarded to him on the basis that the company had abandoned the ground.[6] The Manorburn Sluicing Company then went into liquidation.

With the demise of the Manorburn Sluicing Co Ltd and the death of Rivers in 1910, the weir, built at public expense, became unnecessary. It remained standing for 25 years as a picturesque monument to yet another failed mining enterprise, until it was submerged in 1934 beneath the rising waters of the newly built Lower Manorburn Dam.

The completion of the sludge weir was George Dick's last victory over mining. He died in 1910.

Over the following years the orchard became the main centre of Howden and Moncrieff's activity, as seed production was abandoned and the nursery run down. The place was now called 'Manorburn Orchard.' A crippling blow occurred in the early morning of 23 September 1914 when the homestead, occupied by manager Gadd and his family, was destroyed by fire.

This was not the only problem besetting the firm at this time. George Howden, concerned about the general downturn in business caused by the war, decided to quit the partnership. The lease expired in early 1916 and Moncrieff did not seek to renew it. Instead the property was sold to William Henry Kinraid, who continued to operate it as an orchard for some years. But the gradual raising of the bed of the Manuherikia River by the influx of mining tailings, coupled with the increasing irrigation of the surrounding land, had caused a substantial rise in the water table so that fruitgrowing was no longer possible. The once fine orchard reverted to the lush grassland we see today. Only the lines of tall poplars still mark the boundary of the property where agriculture won an important victory over the mining industry.

NOTES

1. Hale, A. M. 1955 p.88.
2. *Dunstan Times* 1 November 1895.
3. *Dunstan Times* 8 May 1898.
4. *Dunstan Times* 13 November 1898.
5. D'Esterre p. 14-156.
6. *Dunstan Times* 21 March; 30 June; 25 August 1909

18.

SCAM UPON SCAM

— Big Ideas in Little Valley

The road that sidles up the face of the steep ridge directly across the Manuherikia River from Alexandra is a prominent landmark that can be seen from every part of the town. At the top of the hill, behind the large clock set on the hillside, the road takes a sharp U-turn and then runs along the back of the ridge just below the crest. It passes Observation Point before continuing on its tortuous way to Little Valley.

Little Valley is a small basin in the rocky, hilly landscape that stretches southeastwards from Alexandra. To the traveller who has followed the winding, narrow, and in places steep, gravel road for the 12 kilometres from town, the valley with its smooth, irrigated fields, scattered farmsteads and buildings is like an oasis. It is overlooked on the south-east by a long, straight, steep rocky ridge which geologists recognise as a fault scarp, and this gives the clue to the origin of Little Valley — it is a 'fault angle depression.' This means that when a line of weakness developed in the earth, the land to the northwest of it hinged downwards exposing the prominent fault scarp. One effect of this was that the depressed land was protected to some extent from natural erosion, so in the basin is preserved a remnant of the layer of ancient sands and clays which once blanketed much of the Central Otago terrain.

Through the depression runs Little Valley Creek, a tributary of the Manor Burn, which in turn joins the Manuherikia River a kilometre or so above Alexandra. Little Valley Creek[1] and its tributaries (East and West branches, Mt Campbell and Speargrass Creeks) rise to the south of Little Valley amongst the rolling ridges and swampy uplands that are the eastern slopes of the Knobby Range. The main stream enters Little Valley by way of a spectacular gorge cut through the fault scarp.

Little Valley was not always the oasis we see today. Until World War I it was simply part of the vast Galloway Station, only its subdued topography with its rock-free, gentle slopes distinguishing it from the surrounding

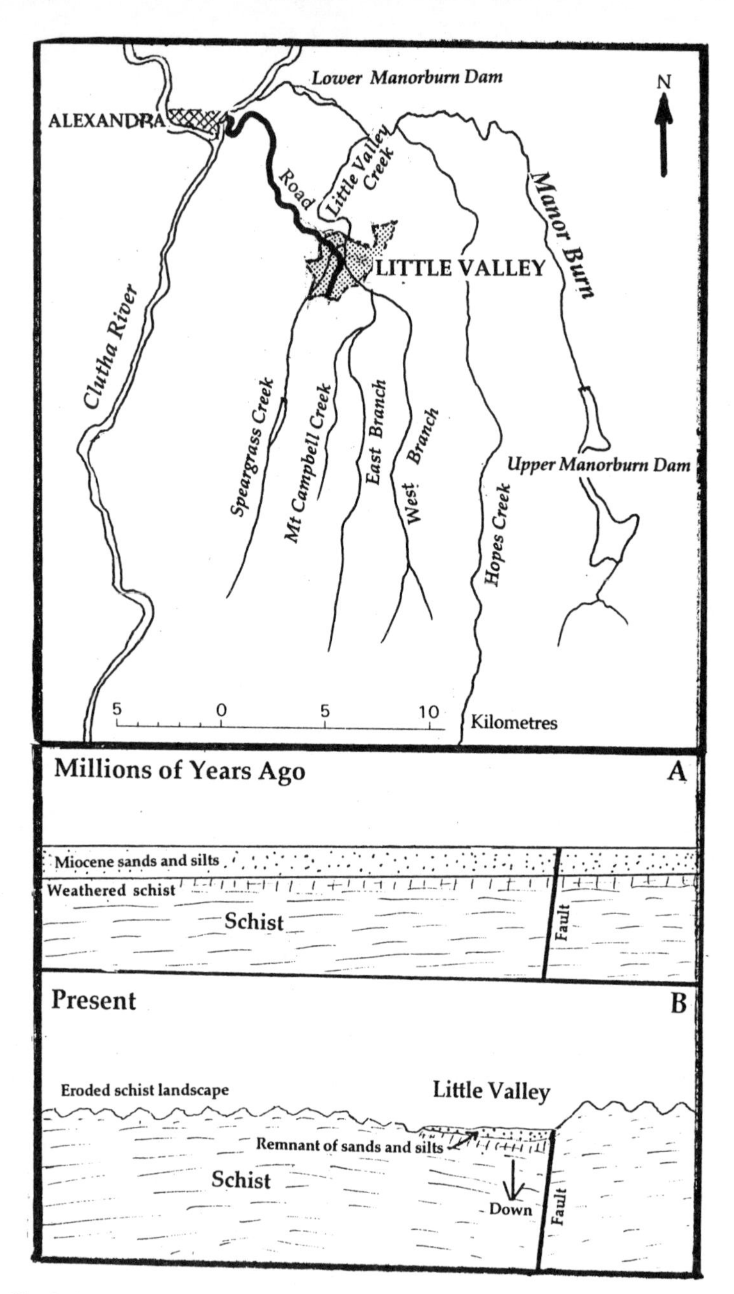

Figure 18. 1. Location and origin of Little Valley ;
Upper: Location map.
Lower: Crosssections showing how Little Valley was formed as a down-faulted depression. A remnant of the layer of old sands and silts that once covered the greater part of Central Otago (A) is preserved after faulting in the depressed area (B).

wilderness of rocky ridges. The occupiers of Galloway Station recognised the Valley as a useful centre for mustering and shearing activities, and an out-station was established with accommodation and large woolshed. The woolshed was destroyed by fire in 1912 with the loss of many sheep and much valuable wool.

There had not been much gold mining in Little Valley in the early years. No doubt prospectors had worked their way through it, as they had through every other corner of Central Otago, but up to 1900 no substantial mining activity had resulted from their efforts. In 1909, however, another search was underway. Two prospecting parties, one consisting of William Berry and Arthur James, and the other headed by William Townsend, were busy testing the Valley and were certainly encouraged, perhaps financially backed, or even employed, by a Mr P. G. M. Fink.

The Little Valley Sluicing Company Ltd

Paul Gustav Maxmillian Fink — you could expect a man with such a name to be a little different from the ordinary gold miner. And so he was. He was an entrepreneur, a man good at forming syndicates of people who were prepared to finance mining enterprises on the chance that they might be successful. He first appeared in Dunedin as Chief Officer of the American barque *Elinor Vernon* in the early 1890s, and a few years later was back to organise a syndicate to work mines at Coolgardie in West Australia.[2] Shortly afterwards he moved to New Zealand and settled in Alexandra.

Fink must have been a modest soul. He gave his occupation for the Alexandra *Roll of Ratepayers* for 1906 as 'dredgeman,' whereas he was actually the dredgemaster and principal shareholder of the *Lady Annie* gold dredge. This dredge worked in the Manuherikia River above the junction of Dip Creek, and although the owners did not publish their gold returns, the fact that the dredge continued to operate for 10 years or more meant that it was successful. But although dredging was Fink's main occupation, it was only one of his interests.

For Fink, 1909 was a particularly busy year. Not only was he interested in the prospecting at Little Valley, but also in opening up an old shaft sunk on a quartz reef between Alexandra and Butchers Gully, and in redeveloping an old mine near Macraes Flat in eastern Otago. He also found time to successfully replace the steam engine on his dredge with an 80 h.p. suction gas engine.

The prospectors soon found ground sufficiently rich to justify pegging out a claim. Berry and James applied for a long narrow strip, of some 80 acres (32 h), running along the foot of the steep scarp that bounds Little Valley to the southeast. At the same time, William Townsend applied for a similar strip of 80 acres lying end-on to that of Berry and James. So between them. these miners held a claim 145 chains (2.9 km) long and 10

Figure 18. 2. The approach to Little Valley from Alexandra. The valley is overlooked on the southeast by a steep range (background) bounded by a fault scarp, which is broken by the rugged gorge of Little Valley Creek.

chains (200 m) wide lying between Little Valley homestead and Hopes Creek. Even before the applications for the claims were finalised the prospectors, calling themselves the 'Little Valley Sluicing Party,' were applying for water rights from nearby streams. Over the next year they were granted 15 heads from the East Branch (including five heads to be diverted into it from Hopes Creek), and another separate water right of 10 heads from Hopes Creek.

Neither Berry nor Townsend had the resources to develop their claims, so let it be known that they hoped a company might be formed to take over their ground. Paul Fink and a party of friends, including an experienced miner, Charlie Weaver, inspected Berry's and Townsend's claims in November 1909. They were apparently impressed to the point where Berry was asked to carry out further testing. Shafts were to be sunk and samples of the wash from the 'bottom' panned off to assess the amount of gold recovered.

After 18 months of testing, Fink was satisfied that the ground contained profitable gold, so he apparently bought the claims and water rights from the prospectors and proceeded to form a public company. The Little Valley Sluicing Company Ltd, was registered on 26 May 1911. The objects of the company were simply stated. The Memorandum of Association said they

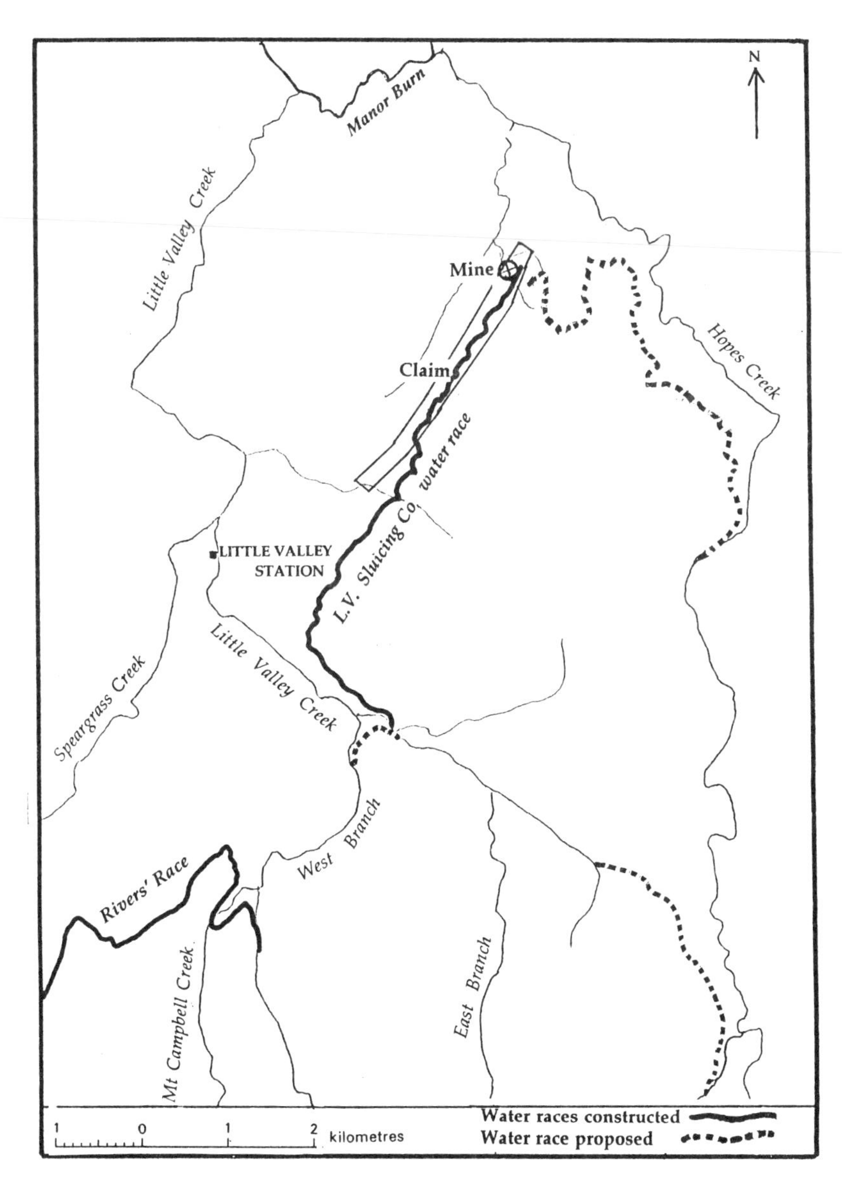

Figure 18. 3. The Claim of the Little Valley Sluicing Company Ltd and the site of the mine. The water race from the East Branch of Little Valley Creek supplied water to the mine. The proposed water races from Hopes Creek and the West Branch of Little Valley Creek are also shown.

were to 'purchase, take over or otherwise acquire' the two alluvial claims and the licenses for four water races.

The company had a nominal capital of £5,000 but only 2,700 £1 shares were allocated, and of these, 1,100 fully paid-up shares were given to

Figure 18. 4. The race (with farm roads above and below) was constructed along the easier hill slopes between the gorge and the mine at a fast pace.

Fink in return for the claim and water rights. Apart from Fink, who was by far the largest shareholder, and whose previous modesty did not prevent him from now describing himself as 'mining engineer,' 15 others held from 50 to 300 shares each. Prominent among them were John Hungerford, gas engineer, and his wife who held 400 shares between them, Walker brothers, plumbers, who took 200 each and P.Y.Wales, the architect, with 200. All the shareholders, apart from Fink, were from Dunedin. Because of the large handout to Fink there was only £1,675 available in cash.

Even before the company had been registered, tenders were called for constructing nine miles (14 km) of water races. The plan was to construct races to divert water from the West Branch and from Hopes Creek[3] into the East Branch, and then to take all of this water by a race over four miles (6 km) long, and capable of carrying nine heads, to the claim.

George Campbell, perhaps the most experienced race-builder in Central Otago, was appointed as supervisor. A camp was established near the Little Valley homestead and in July 1911, 26 men started construction. The greater part of the race was in what was described as 'easy country,' and progress here was rapid, so some three miles were completed by Christmas 1911. The route up into the Little Valley Creek gorge, where the intake of the race was situated, was a different story. Here there were steep rocky hillsides that required hundreds of yards of high stone walls to support the race, but worst of all was what Campbell described as 'the bad half-mile.' This was a long stretch of precipitous

Figure 18. 5. There was great difficulty in bringing the Little Valley Sluicing Company's race out of the gorge of Little Valley Creek. Long lengths of stone walling were needed to support the race along the steep, rocky hillside.

broken rocky hillside that required more than 800 feet (250 m) of fluming.

It was during the summer of 1911 that the company realised just how little water these creeks carried. It was one thing to be granted a water right of 15 heads by the Warden's Court, but the warden didn't supply the water — he merely gave the right to take it if it were available. And for the greater part of the year it wasn't available. So the quest for even more water went on. Another nine heads from Little Valley Creek were granted in 1912, and an attempt was made to acquire the 10 heads from the West Branch and Mt Campbell Creek (always referred to as the 'Mt Campbell water rights') held by the Rivers' Estate but controlled by the Alexandra Borough Council. The overtures were refused.

It took most of 1912 to complete the main race, which had been estimated to cost £2,000 but was claimed to have eventually cost £4,700. It was so expensive, in fact, that construction of the other proposed races from the West Branch and from Hopes Creek was postponed until the mine had been proven. Finally at the end of the year water arrived at the claim.

Sluicing commenced immediately, and after a few weeks the 'wash-up' took place. This event was always looked forward to with eager

Figure 18. 6. An air view of the claim shows that only a very small amount of sluicing was done before the mine was abandoned.

expectation by miners, but in this case it resulted in disappointment — there was little sign of gold[4] when the mats in the sluice boxes were cleaned down. There was an urgent meeting of the shareholders. Money had already been borrowed from the bank to complete the water race, so there was little choice but to go into liquidation and offer the company for sale. The pipes and other mining gear were leased for 10 shillings a week to the Doctors Point Mining Company Ltd, which carted them over to its claim high up on the wall of the Clutha River gorge about eight kilometres below Alexandra.

In December 1913,[5] the Mining Registrar successfully proceeded against the company for 'forfeiture of mining privileges' and judgement for the amount of rent due. In other words, the claim was taken back by the Crown but the defunct company still had to pay the back rent. And that, apparently, was the end of the Little Valley Sluicing Company Ltd, but it is important to note that the various water rights were not forfeited. A group of the principal shareholders (but not including Fink), who were obviously looking to the future, applied for and gained 'protection' of the water rights through the Warden's Court. While under 'Protection' the races could not be classed as 'Abandoned,' and so taken over by someone else.

A small gash in the land below the end of the water race, today marks the site where one of the Big Ideas for Little Valley, to say nothing of the shareholders' hopes and money, were washed away.

What had gone wrong? Why had ground, which allegedly had been thoroughly tested with good results yielded little or no gold? In cases such as this, rumours of 'salting'[6] are always rife. By this is meant that gold from some other locality is 'planted' in the claim to deceive prospective buyers into thinking that the claim is much richer than it really is. A favourite method apparently was to load a shotgun with gold and blast it into the ground. The Little Valley claim was no exception to such rumours, which are still heard to this day, but proof of such fraud is nearly always impossible to establish.

The only ones who made money out of the claims were the prospectors who sold them to Fink on the basis of their rich prospects.

The principal Dunedin shareholders were particularly upset. Not only had they lost the value of their shares, but they had also guaranteed money borrowed from the bank. Their thoughts turned to ways and means of recouping some of the money they had lost in the ill-fated enterprise.

The Alexandra Development Party

Public meetings were a feature of life in Alexandra. They were called by successive mayors to discuss such diverse matters as the vexatious water supply; fund-raising for patriotic purposes; a new fire station; electric power supply; progress of irrigation schemes and many other matters of public interest. And, in the days before television and other distractions, they were well attended. They gave the councillors a chance to demonstrate their continued interest in the progress of the town and to show off their skills at oratory. Members of the public enjoyed criticising councillors and each other.

On the evening of Friday 13 February 1914, a Mr George Neill was in full cry at such a meeting. He was outlining a proposal to subdivide and develop Little Valley, a part of Galloway Run. It is a fair bet that few of the audience knew where Little Valley was, let alone had been there, but they were sympathetic because mining was fast declining and there was a great demand for small land holdings.

Who was this Mr Neill? He was a Dunedin land agent who had already persuaded a group of wealthy Dunedin businessmen that they should buy a large farm near Cromwell, subdivide it into small blocks, install an irrigation scheme then sell the blocks for fruitgrowing. Many aspiring orchardists would thus see their dreams come to fruition, the district would benefit greatly, the Dunedin syndicate would make a handsome profit, Mr Neill would collect his commission with each transaction and everyone would be happy.

Now, as he addressed his Alexandra audience, the Cromwell scheme, still in its early stages, was proceeding smoothly, and Mr Neill's star was shining brightly. Fortunately no one could foresee the problems ahead for the syndicate, which eventually became the Cromwell Development Company. Difficulties in damming the Kawarau River and the eventual destruction of the pumping system by floods led to financial failure, with the flourishing orchards of Ripponvale left to find alternative irrigation water.

Influential men behind the Cromwell Development Company had persuaded Government to pass a special Act, which allowed companies to set up private irrigation schemes, but only in Vincent County. Neill explained that a group of his Dunedin businessmen friends intended to make use of this legislation to set up a development scheme in Little Valley.

The Act

Government had passed, in December 1913, the Water-Supply Amendment Act. Section 6 of this Act was of great importance to Central Otago. It said, in summary, that if a company were set up with the intention of irrigating, subdividing and selling land for settlement within Vincent County, then the Government would, under certain conditions, sell Crown land to the company for this purpose. Only in exceptional circumstances (such as the need to have a title in order to borrow money) would the company itself receive a title to the land—titles were to be issued directly to the settlers to whom the company resold the land. In other words it was a land development company. It would buy undeveloped Crown land, irrigate it, subdivide it and sell it off to potential farmers at a substantial profit.

The conditions under which the land was made available to the company were not unreasonable: the land had to be irrigated according to approved plans and methods, the land had to be surveyed and subdivided within a certain time, the subdivisions were not to exceed a certain size and no settler was to be allowed to acquire more than one property.

The Little Valley Scheme

Neill pointed out to his audience, at two public meetings,[7] that this legislation provided opportunities for land development and settlement too good to ignore. The Dunedin syndicate had been formed with the intention of buying a block of land in Little Valley from the Crown under this legislation. Then, using the water race and the water rights of the defunct Little Valley Sluicing Company, and a storage dam that was yet to be built, it would supply irrigation water to a large number of small landholders. The thrust of Neill's address was to persuade the people of Alexandra to support the scheme, not necessarily financially, but by persuading Government to separate Little Valley from the Galloway Run and to subdivide it into small sections. The lease on Galloway Run expired in 1916 and there was an expectation that the Run would be cut up then into Small Grazing Runs. But Neill wanted pressure applied to make sure that Little Valley was not allocated to one or more of the Small Grazing Runs. He wanted it separated and cut up into small units.

A sympathetic Alexandra audience passed resolutions pledging support, and also asking the Prime Minister and the member for the district to do what they could to assist the syndicate to carry out its objectives.

Irrigation

The possibilities of irrigation for Little Valley had already been investigated by the Public Works Department. After James Rivers's death in 1910, his Trustees offered to the Government the Mt Campbell water rights, the Speargrass Creek dam and the long race from the dam out to Galloway. £2,000 was the asking price. District Engineer, F. Furkert, estimated (without a survey) that there might be 1,900 acres of arable land in Little Valley. But he pointed out that, even allowing for the fact that not all of this land would be irrigable, the amount of water available in the Rivers scheme would irrigate only a small fraction. And besides, the price was far too high, so it was recommended that the offer be turned down.[8]

Borough Sells Water

Back in August 1902 the Alexandra Borough Council had signed an agreement with James Rivers that, for the sum of £1,200, the council would acquire a one tenth interest in Rivers' water rights, dams and races. As Rivers claimed he had rights to 10 heads of water from Speargrass Creek and the Mt Campbell water rights, the council assumed that it was entitled to a supply of one head for town supply. It was not to be. As it turned out, Rivers did not have 'prior right' to the 10 heads, and this fact, coupled with dry seasons and a long race that was difficult to maintain, meant that the Borough was constantly short of water. In the end the Borough, in spite of much litigation, was forced to acquire a new water supply from Butchers Creek, and the water from Rivers' race was leased out to miners.

Peter Walker, the Secretary of the Little Valley Sluicing Company, had tried to buy Council's head of water but was refused. Now a fresh approach was made by George Neill on behalf of the Alexandra Development Party, as the Dunedin syndicate was now calling itself. This time it was successful. In July 1914 the Alexandra Borough Council resolved to sell its interest in Rivers' water to the Party for £500. It took nearly 18 months for the agreement to be hammered out, but finally in November 1915 the documents were ready for signing, but it still took another couple of months and a threat to withdraw from the deal before the cheque was paid.

At the same time the Party agreed to buy from the Rivers Estate the remaining interest in the Speargrass Creek Dam and the Mt Campbell water rights. The price was £2,000, but it was agreed that the Party would take over the mortgage of £720 held by the Trustee of Stephen Foxwell's estate, and the Trustees of River's estate accepted a second mortgage over the remainder. The syndicate now owned the water rights for nearly all available water, some 60 heads in all. Apart from a water right owned by Robert Campbell and Sons for the use of Galloway Station, the Development Party now had control of all of the water in Little Valley, water which might be required for future irrigation of the Valley and for stock water.

It might be noted in passing that the only cash exchanged so far was the £500 to the Alexandra Borough Council, and that was apparently paid only with great reluctance.

The Alexandra Development Party now owned the following water rights:-

No 1396	5 heads from West Branch
No 1405	5 heads from Hopes Creek
No 1453	10 heads from Hopes Creek
No 1515	15 heads from West Branch
No 1700	9 heads from East Branch
No 1736	6 heads from West branch
	10 heads Rivers' and Alexandra Borough Council's Mt Campbell and Speargrass Creek water rights

The total was 60 heads. What was the Party going to do with this large amount of water?

Perhaps the first public indication of what was in the wind was when James Clark, Robert Rutherford and John Hungerford, all of whom had been shareholders in the defunct Little Valley Sluicing Company, applied to the Warden's Court in April 1914 for a 'Change of Purpose' of the water rights that had been held by the old company. They wanted the purpose changed from 'Mining' to 'Irrigation and Domestic.' Having been granted this they promptly applied for further 'Protection.' Then the same three shareholders each applied for, and was granted, a 'Residence Area' of one acre in Little Valley.

Full of confidence, the Party's solicitor wrote to the Commissioner of Crown Lands[10] asking, under the 1913 Act, for 2,000 acres to be made available to the Party for irrigation when the Galloway Run was broken up in 1916. If the Land Board, which had jurisdiction in such matters, didn't agree with this proposition, said the solicitor, then the Government could buy the Party's water rights and races, for the bargain price of £8,500.

The Commissioner advised the Minister of Lands that this price, plus the cost of raising Rivers' dam by 15 ft, would mean a financial loading of £3 on every irrigated acre. This was not to be considered.

The Alexandra Development Party Ltd

Everything came out in the open when the Alexandra Development Party Ltd was registered as a public company in September 1915. It was listed as having £8,000 capital, and 5,700 £1 shares were allocated to 14 shareholders.

It came as no surprise to some to find that the principal shareholders had been the principal shareholders in the old Little Valley Sluicing Company. Certainly Paul Fink, the mining engineer was missing from the list, but John Hungerford, Robert Rutherford, John Clark (by then Mayor of Dunedin), Peter and James Walker and Patrick Wales had all been shareholders in the old Sluicing Company.[11]

The objectives of the new company took eight pages to list, and covered everything from irrigating, subdividing and selling land to generating electricity, running saw mills and coal mining.

The main concern, however, was clearly with fruit. The company apparently planned to grow it, cool it, sell it, export it, preserve it, make jam with it, can it, and deal in it.

Fruitgrowing

The arrival of the railway at Clyde in 1907 had caused a huge boost to the fruitgrowing industry. For the first time fruit could be delivered to Dunedin and other markets quickly and in good order. Fruitgrowing had captured the imagination of Dunedin businessmen, and a number of large fruitgrowing schemes were instigated. First it was the Fruitlands Estate at Bald Hill Flat, followed by the Cromwell Development Company's scheme at Ripponvale, then the Terrace Orchard Company at Alexandra. Why not fruitgrowing at Little Valley when the Galloway Run was broken up ?

Mr Blackmore, the Government 'Pomologist,' had already reported[9] to the Alexandra Progressive League on the fruitgrowing potential of Little Valley. .In many ways the report was similar to that which Blackmore had written about Bald Hill Flat a year or so before. Everything was in favour of fruitgrowing—topography, soil, climate, access to rail and so on. All that was needed was an adequate and reliable supply of irrigation water, and to this end he strongly recommended the League approach Government to urge it to buy Rivers' water rights before it was too late

Cut-up of Galloway Run

No doubt because of Pomologist Blackmore's glowing report, the idea spread that Little Valley would be ideal for irrigated fruit farming. A small area of alluvial soils in the centre of the valley was already freehold land owned by Galloway Station, but it was envisaged that the roughly horseshoe-shaped area of easy sloping foothills that almost surrounded the central flat land, would be ideal for orchards.

At a meeting[12] of the recently formed Alexandra branch of the Otago Expansion League, however, not everyone was in favour of close subdivision. There were those, farmers no doubt, who thought that separating off this fertile, potentially irrigable land from the small runs that were to be formed out of the Galloway Run, would make their working more difficult and reduce their value.

This ambivalence engendered an editorial from the local newspaper.[13] It reminded the Borough Council that, knowing full well the objectives of the Alexandra Development Party Ltd, it had been more than willing to unload its otherwise unsaleable water right on to the Party for the not inconsiderable sum of £500. In the editor's view the council and town now had a duty to fully support the Party in its endeavours to have Little Valley subdivided and opened up for settlement.

In June 1916, Galloway was cut up into five Small Grazing Runs in such a way that three of these new runs could have homesteads on the freehold land that the Government had purchased in Little Valley.

Now the officials began to apply pressure to break the monopoly that the Development Party held on Little Valley's water. First, the Commissioner of Crown Lands asked the Warden's Court to forfeit the Company's water rights because they were not being used. The Company really had no defence and could only say it was irrigating the Residence Areas. The thought of 60 heads of water being used to irrigate three blocks of one acre each must have caused some merriment in the court. But the pointed comment by the Company that the water was meant to irrigate 2,000 acres was a jab at the Land Board, which had recently turned down the Company's request for this large area. Nevertheless, it looked as if forfeiture of the water rights was inevitable, although William Bodkin, the Company solicitor and a shareholder, managed to get the case adjourned for a month. At the same court session, however, the warden, at the request of the Mining Registrar, tightened the screw by forfeiting the three Residence Areas held by the three principal shareholders, as they were not being used and the rent had not been paid.[14] Now the company had no excuse at all for holding on to water for irrigation. Unless it moved quickly the water would be forfeited to the Crown.

Politicians Take a Hand

In desperation James Walker, the Company secretary, appealed to the Prime Minister, Mr Massey:

> Sir
>
> I am instructed to offer to the Government all of the Company's water rights on Galloway. We would have preferred to take up land and still

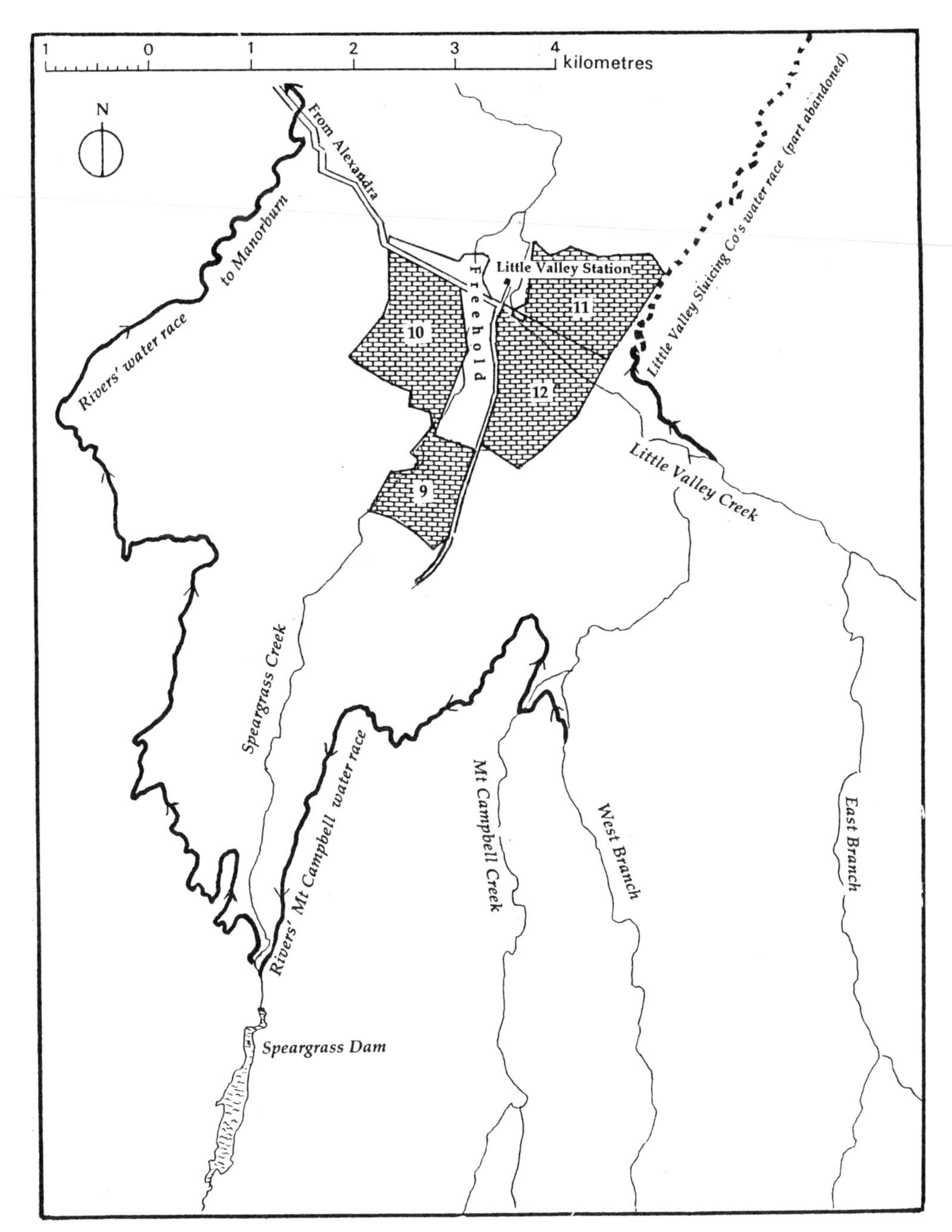

Figure 18. 7. The 888 acres of land allotted to the Alexandra Development Party Ltd was surveyed into four sections in 1917. Irrigation water was supplied from a renovated part of the old Little Valley Sluicing Co's race, and from part of the long Rivers' race which once supplied water to mines at Galloway from the Mt Campbell race and from Speargrass Dam.

prefer to do so but the Land Board is not prepared to favourably consider such a proposal. The water rights actually cost over £9,000 but we have decided to accept the sum of £4,500. [15]

The letter had an effect. At the next meeting of the Land Board, held on 12 July 1916, 900 acres (later confirmed by survey as 888 acres) at Little Valley were set aside for the use of the Alexandra Development Party Ltd.[16]

An Agreement[17] between the Company and the Crown was drawn up in July 1917. The main points were that the Company was to:-

pay £1,567 15s. 5d. (£1 12s. 6d. per acre) for the land and £126 13s. for existing fencing.
irrigate the land and fence it within one year,
sow 200 acres in lucerne or fruit trees within two years,
subdivide the land and offer it for sale or lease with a purchasing clause,
supply water at the rate of half a head for each 100 acres,
request the Crown to offer *bona fide* purchasers a Certificate of Title.

In anticipation of the Agreement being signed, Lands and Survey Department surveyed off the allotted land and cut it into four roughly equally sized sections. The Company took a long time to consider the Agreement and asked for several changes, mainly to do with the proposed requirements for subdivision, pointing out the impossibility of obtaining wire netting during wartime. It was still considering when there was a dramatic development.

The Whistle Blower

William Bodkin, a solicitor in Alexandra who was very interested in the development of the district, held 225 shares in the Alexandra Development Party Ltd and acted as their solicitor. On 30 October 1917 he walked into the office of the District Commissioner of Crown Lands in Dunedin and made a statement.

Bodkin said that it was he who had arranged the public meetings in Alexandra and was responsible for statements regarding the *bona fides* of the company. He wished to say that he was thoroughly dissatisfied with the proceedings and wished to make a clean breast of the business. It was his opinion that the Company was holding the land simply as a speculation and had no intention of developing it. The Company was merely trying to attract people to put in money to reimburse it for losses in connection with the water rights. He pointed out that apart from the £500 paid to the Alexandra Borough Council, the Company had not put one penny into development of the scheme. He finished by saying how glad he was to have been able to get this statement off this mind.[17]

The District Commissioner reacted quickly. All dealings with the Company were suspended while he made inquiries. He confirmed that the

Figure 18. 8. William (later Sir William) Bodkin, although the solicitor to the Company and a shareholder, felt he should expose the true intentions of the Company promoters to the Government.

original promoters of the Company, who owned the water rights, had not contributed anything to the venture. The £500 paid to the Alexandra Borough Council was obtained from several small and more recent shareholders. Furthermore, these promoters had allotted to themselves 5,000 fully paid up shares so that if the water rights were sold, they would recoup their losses in building the Little Valley Sluicing Co's water race, on which £800 was still owed to the bank. The Commissioner was of the opinion that the Company was simply encouraging people to put in money that would be used to reimburse the original shareholders for that which they had sunk in the water rights and lost on mining. He concluded in his report to the Under-Secretary:

> I feel sure that the Alexandra Development Party comprised speculators who had no intention of fruit farming but who made it a business to secure a monopoly over water with a view to unloading at the expense of the legitimate farmer.[18]

The Under-Secretary responded by asking for a Statement of Affairs from the Company. When this was produced, it confirmed that of the 5,700 shares issued, 5,000 fully paid up shares had been given to the major shareholders, who were also the promoters, for the Little Valley water rights. It was the other contributing shareholders who provided the cash to buy the Alexandra Borough Council water right. A mortgage had been arranged to allow the purchase of the Rivers' water rights and dam. There was £2,300 of unallotted capital, which was being kept back to pay the Crown for the land in Little Valley.

The Company finally signed the Agreement on 21 March 1918, and it paid over the £126 13s to cover the value of fencing already on the block. But still the company made no move towards further subdivision or irrigation of the land.

In February 1919, the Public Works Department applied for forfeiture of most of the very large quantity of water over which the Development Party held rights. Evidence was given in the Warden's Court that the promised races had not been constructed, and those that were had not been used for a long time. Fluming was down and pipes missing. The Party admitted that it had had difficulties, and had not used the water rights but had regularly applied to have them protected. Fencing materials, especially wire netting, could not be obtained owing to the War, and subdivision could not be proceeded with. The Land Board had, indeed, waived the special conditions attached to the grant. In the end the warden forfeited all of the Development Party's water rights except for four heads—two from the East Branch and two from the West Branch.[19] These were deemed more than sufficient to irrigate the 360 acres, which was all of the Development Party's area that could be commanded by water. The remainder of the Party's water was forfeited to the Crown.

As late as August 1919, the Company was asking for yet another 12 months extension of the Agreement, pleading the usual excuse of difficulty in getting wire netting, etc. This time it promised faithfully that, if the extension were given, it would then pay for the land and obtain a title. Already George Rivers had been instructed to start renovating the race from Speargrass Dam and begin fencing.

The Company's original plan, to recoup the shareholders losses in the mining venture by selling their water rights to the Government, had failed. And now the Company was constantly finding excuses for not going ahead with its back-up plan of subdividing the land into small, irrigated fruitgrowing blocks for resale to settlers — if it ever had any serious intention of doing this. The principal shareholders were simply not prepared to spend money on the subdivision fencing, irrigation races, etc. necessary to meet the terms of the Agreement. Now they had lost, by forfeiture, the bulk of the water that they saw as their main weapon in their battle to recoup their money. It was time to get out and salvage something from the shambles. It was decided to sell off the land, not as small fruitgrowing sections but as four blocks suitable for farming.

In mid-1919 George Rivers, a storekeeper in Alexandra and a spokesman for the Trustees of the Rivers' Estate, which still held the mortgage on the Mt Campbell water rights, decided to retire from his

Figure 18. 9. An air view of Little Valley looking east. The clump of trees (right foreground) marks the homestead of the only farm sold by the Alexandra Development Company under the 1913 Act. It was bought in the name of Mrs George Rivers and developed as a dairy farm.

store. Shortly afterwards he turned up in Little Valley repairing the Speargrass Creek water race, presumably under some kind of arrangement with the Company. In May 1921 Mary Rivers, wife of George, was granted a Title to Section 9 on the recommendation of the Development Company. This was the only section disposed of in the manner prescribed by the Water Supply Act of 1913.

After at last paying the agreed price of £1,442 13s 10d for the land, the Company itself applied for a Title for the remaining three sections and this was granted in February 1922. In August 1924 these sections were sold[21] and in early 1925 the company went into voluntary liquidation.

The £2,936 received from the sale of the sections covered the price paid for the land, paid off the bank, most of the first mortgage on Rivers' water rights and other sundry expenses, but left only £14 0s 4d in the bank. Once again the shareholders were out of luck and had, at last, to face the fact that they had finally lost their money.

When the Company was liquidated, the Mt Campbell water rights, which included two heads from the West Branch of Little Valley Creek, the Mt Campbell water race, the Speargrass Creek dam and the water race from the dam, reverted to the Rivers Estate that held the mortgage. The water was put to good use in irrigating the dairy farm George and his successors ran for many years, and who were able to boast that they had the only freehold water rights in Central Otago.

NOTES

1. About 1.5 km above the mouth of the gorge, Little Valley Creek divides into the East Branch and West Branch. Formerly the East Branch was called 'Bikerstaff Creek' and the West Branch was called 'Mt Campbell Creek'. The name 'Mt Campbell Creek.' is now applied to a tributary of the West Branch.
2. *Otago Witness* 16 February 1897.
3. There is no evidence that either of the 'feeder' races from Hopes Creek or from the West Branch was ever constructed.
4. *AJHR* 1913 (covering 1912). Results were 'not too encouraging.'
5. *Alexandra Herald* 17/12/1913.
6. Salting consisted of secretly placing gold in a claim so that it would be found by unsuspecting prospective purchasers. It is said that a favourite method was to blast gold particles into the ground from a shotgun.
7. *ibid.* Reports of meetings 18 February 1914 and 25 March 1914.
8. F. W. Furkert to Under-Secretary Public Works Dept. 24 July 1912. File D 84 1 0/3/7 Regional Office, National Archives, Dunedin.
9. *Alexandra Herald* 19 January 1913.
10. R. M. Rutherford to Commissioner of Crown lands Dunedin. 5 August 1914 File D 84 10/3/7 Regional Office, National Archives, Dunedin
11. The Principal shareholders in each company were:

Little Valley Sluicing Co and Alexandra Development Party Ltd
P. G. M. Fink, mining engineer
John Hungerford, gas engineer
Robert Rutherford, company manager
James Walker, sharebroker
Peter Walker, plumber
George Neill, estate Agent
John Clark, Gentleman

12. *Alexandra Herald* 8 March 1916.
13. *ibid.* 22 March 1916.
14. *ibid.* 10 May 1916.
15. J. H. Walker to Prime Minister 2 June 1916 File D 84 10/3/7 Regional Office, National Archives, Dunedin.
16. Otago Land Board Minutes of Meeting 12 July 1916. Reported *Otago Witness* 14 July 1916.
17. File D 84 10/3/7 Regional Office, National Archives, Dunedin.
18. *ibid.*
19. Commissioner of Crown lands Dunedin to Under-Secretary for Lands 2 November 1917 File D 84.10/3/7 Regional Office, National Archives, Dunedin.
20. *Alexandra Herald* 19 March 1919.

21. The four sections were sold as follows:

No. 9 Mary Rivers

No. 10 Alexander McKenzie

No. 11 Patrick Ritchie

No. 12 Peter Wallace Curtis

ACKNOWLEDGEMENTS

I am grateful to many people and organisations who have helped assemble the information in this book. The University of Waikato, particularly Campus Photography and the Department of Earth Sciences, continues to provide steady support. The Interloan Department of the University Library has over many years arranged a stream of theses, scarce books, and reels of microfilm of early Goldfield's newspapers.

Hocken Library and the Dunedin office of New Zealand Archives have continued to be most helpful in searching out photographs and information.

People who have shown me sites and often provided off-road transport, included Mark Reid, Peter Dunbier and Lyndon Sanders. Others who provided information or photographs include Ernie King, Jean Blewett, John and Betty Wilson, George Elder, John Symons all of Alexandra, Syd Attfield of Clyde, Wayne Stark of Christchurch, George Griffiths of Dunedin and Noel Watts of Cambridge.

I am grateful to the President, Gill Grant, and the Committee of the Alexandra District Historical Association for arranging the facilities for book launchings and allowing me to copy photographs from the photographic library of the William Bodkin Museum. Former Museum managers Helena Heydelaar and Cheryl Grubb, and the present manager Elaine Gough, have been very helpful in providing information and facilities. I am particularly grateful to Joan Stevens of Alexandra who has spent much time in trying to find answers to difficult questions about people and events of the early days.

I thank my publisher, Trevor Reeves, for his patience and help and his wife, artist Judith Wolfe, who provides the illustrations for the covers of the books.

Finally, grateful acknowledgement is made to Dr David McCraw for field companionship, field assistance and editing and to Jean King for skilled proof reading.

Acknowledgement of Photographs

Alexandra William Bodkin Museum Figures 1.3; 1.13; 4.4; 4.7; 4.8; 4.10; 6.10; 7.7; 9.1; 10.1; 11.7; 17.4; 18.8

Anderson, G. A. Figure 8.1

Elder, G. Figures 13.4; 15.6

Hocken Library Figures 3.8; 5.2; 7.2; 17.3; 18.5

Lands and Survey Department Figures 7.3; 8.1; 11.11. 17.5

Lunn, Bert Figure 16.9 (upper)

McCraw, Dr D. J. Figure 2.8

Butler, M. of NIWA Alexandra Office Figures 5. 4; 5.5

Symons, J, 16.6

All other photographs were taken by the author or are part of the author's collection from various sources and accumulated over many years.

GLOSSARY

Adit: strictly, the entrance to a horizontal passage excavated into a hillside but often used for the whole passage which is strictly a 'drive.'
Bottom: bedrock on which the 'wash' lies. A hard layer within an alluvial sequence on which gold accumulates, is a 'false bottom.'
Battery: see 'stamper-battery'.
C-dam: a small dam with a low, curved earth wall (hence the name) built across a shallow gully or depression in which water accumulated overnight to allow sluicing during the following day.
Chair: a seat attached to a rope stretched across a river in such a way that the occupant could pull the seat along the rope.
Cradle: a small portable box fitted with gold saving slats, expanded metal and matting on a series of sloping shelves. 'Wash' placed in the top of the cradle was washed through the device by water, assisted by rocking of the cradle. Every so often the mats were removed and shaken into a gold pan and the gold recovered.
Drive: a more or less horizontal passage driven into a hillside. (cf 'tunnel')
Fault: a fracture in the earth's surface allowing the land on one side of the fracture to rise or sink relative to the other. This movement gives rise to a 'fault scarp'—a straight, steep face which is characteristic of many ranges in Central Otago.
Ground sluicing: a ditch lined on the bottom and along the sides with flat stones which trapped gold separated from the 'wash' shovelled into the ditch by water running through the ditch. The debris passed out of the ditch as 'tailings.' During a 'wash up' the stones were removed, the trapped black sand and gold shovelled and swept out and the gold separated from the sand by panning or cradling.
Head: abbreviation for 'Government sluice-head.' Water flowing at the rate of one cubic foot per second. Miners were allocated water on the basis of one sluice head for a claim supporting one or two miners, two sluice heads for four miners and so on.
Hydraulic elevator: a device for lifting gravel, sand and gold out of a deep mine into sluice boxes at a higher level where tailings could be disposed

of. The necessary suction was created by a water jet blasted up a vertical pipe.

Hydraulic sluicing: the removal of auriferous ground by a jet of water under high pressure directed from a nozzle.

Lode: (from *lead*) a tabular deposit of valuable mineral between definite boundaries. Synonymous to some extent, with 'vein' and 'reef.'

Moraine: a characteristic hummocky landform formed from rock debris deposited from a glacier when the ice melts.

Mullock: soft, non gold-bearing rock associated with quartz reefs. In Otago, mainly crushed schist.

Paddock: a claim was divided into small blocks or 'paddocks' for ease of working and these were excavated systematically.

Quartz: white, hard mineral comprised of silica (silicon dioxide), often crystalline in appearance, which is abundant as veins and streaks in schist rock. It often contains gold.

Pan: a tin basin with sloping sides. A shovel-full of wash is placed in the pan, and with one edge of the pan under water, the pan is swirled in a gentle circular motion until the debris is removed by the water leaving, hopefully, any gold in the pan.

Reef: so-called because of the resemblance between a lode of hard quartz protruding above the surface and a rock reef protruding from the sea. 'Reef' was also used for 'bedrock' as in 'They worked down until they reached the reef.' Many of the so-called 'reefs' or lodes on the Old Man Range were not quartz reefs but seams of crushed and decomposed rock.

Salting: the fraudulent placing of gold into a mine or sample to make it appear richer than it was.

Schist: the basement rock of Central Otago. It has a layered texture like the leaves of a book, emphasised by white layers of quartz between layers of dark minerals. Mica is normally abundant. Schist is a metamorphic rock, that is, a rock that has been changed from the original fine sandstone (greywacke) by heat and pressure through burial deep within the earth.

Sluice: a device in which flowing water is used to separate gold from the wash Early sluices were a ditch lined with flat stones. Later, sluice boxes were used.

Sluice boxes: long, inclined wooden troughs with strips of wood or iron fastened across the floor to trap gold. Later expanded metal, baize, or matting was used as well.

Stamper-battery: a machine driven by a water wheel, pelton wheel, or steam engine which consisted of a number of vertical rods, each with a heavy iron weight attached, which were lifted and allowed to fall in sequence on broken 'stone' crushing it and releasing any gold.

Tailings: the gold-free debris consisting of sand, gravel etc, discharged from a sluice after the gold had been separated.

Tail race: a ditch that carried tailings away from the sluice. The tailrace had to have sufficient fall for the tailings to be carried along by the available water flow. Some tailraces were a kilometre or more long and some were communal facilities serving several claims.

Tertiary: The period in Geological Time from 2 millions years ago to 65 million years ago. For part of this Period most of Central Otago was covered by a lake and the sediments deposited in this lake are still preserved under the gravels of the present day valleys. They are well exposed as cliffs just above the Galloway bridge.

Tor: a freestanding column or block of schist rock rising from a more or less flat rock platform. Tors range in height from less than a metre to the 23 metres of the Obelisk on the Old Man Range. Should be distinguished from irregular 'outcrops' of rock which are common in stream valleys and on steep slopes.

Tunnel: an underground passage opening to the surface at both ends (as in a railway tunnel). The 'tunnels' driven into hillsides in search of gold are technically 'drives' but were commonly described as 'tunnels' in newspaper reports of the time

Wash: short for 'wash-dirt,' also 'called 'pay-dirt.' A layer of sand and gravel commonly only 30 cms or so thick, which rested on the 'bottom' and contained the gold.

References

Official Documents

Wardens' Court Records: Alexandra, Clyde, Roxburgh. Regional Office National Archives, Dunedin

Company records ,Regional Office National Archives, Dunedin

Otago Provincial Gazette

Votes and Proceedings Otago Provincial Council

Ordinances Otago Provincial Council

Appendices to the Journal of the House of Representatives (AJHR)

New Zealand Statutes

New Zealand Mines Record.

Newspapers:

Otago Daily Times

Otago Witness

Dunstan Times

Alexandra Herald

Cromwell Argus

Tuapeka Times

Mt Benger Mail

Books and Journals

Angus, J. H. *One Hundred Years of Vincent County* Dunedin 1977.

Armstrong. J. F. *On the Naturalised Plants of the Province of Canterbury.* Transactions and Proceedings of the New Zealand Institute, Vol. IV pp. 284-290 Wellington 1872.

Barry, J. *Glimpses of the Australian Colonies and New Zealand* Auckland 1903.

Bodkin, A.W. *An Account of Monte Christo* Clyde 1988.

Bruce, J. L. Dobson. J. H. *Irrigation in Central Otago* AJH C—17 1909.

Chandler, Peter *Glenaray* Invercargill 1984.

Cyclopaedia of New Zealand Vol IV *Otago and Southland* Christchurch 1905.

D'Esterre, Ernest *Central Otago: Its Prospects and Resources.* Dunedin 1903.

Don, Alexander *Annual Inland Tour — Report* The Christian Outlook July 5 1897.

Galvin, P. *The New Zealand Mining Handbook.* Wellington 1906.

Gilkison, R. *Early Days in Central Otago.* Dunedin 2nd Ed. 1936.

Glasson, H. A. *The Golden Cobweb.* Dunedin 1957.

Grey, J. Grattan *Australia, Tasmania, and New Zealand from Earliest Times* Melbourne n.d. (c 1927).

Hale, A. M. *Pioneer Nurserymen of New Zealand* Wellington 1955.

Hall-Jones, John *Goldfields of the South* Invercargill 1982.

Hamel, G. E. *Gold Mining and farming in the Upper Waikaia, an archaeological survey.* Unpublished Report for Department of Conservation 1989.

The Hidden Goldfields of Earnscleugh Unpublished Report for Department of Conservation 1994.

Hannah, Myra *Operation Waterwheel* Alexandra n.d (c 1975).

Hassing G. M. *The Memory Log of G. M. Hassing* Dunedin 1911.

Lovell-Smith, E. M. *Old Coaching Days: Otago and Southland.* Christchurch 1931.

McCraw, J. D. *Mountain Water and River Gold* Dunedin 2000.

The Golden Junction Dunedin 2002

Miller, F. W. G. *The History of Waikaia*: Dunedin 1966.

Moore, C. S. W. *The Dunstan* Dunedin 1953.

Park J. *On the Galloway Alluvial Gold-Diggings* Reports of Geological Explorations 1888-1889 Wellington 1890.

On White's Reefs Alexandra .Reports of Geological Explorations 1888-1889 Wellington 1890.

The Geology of the Area covered by the Alexandra Sheet, Central Otago Division. New Zealand Geological Survey Bulletin No. 2 1906.

Pyke, Vincent *History of the Early Gold Discoveries in Otago* Dunedin 1887.

Salmon, J. H. M *History of Gold Mining in New Zealand* Wellington 1963.
Stevens, Syd *I'll be Damned.* Rotorua 1988.
Stone, Jenny *Archaeology and History of Chamouni* Unpublished
Dissertation Department of Anthropology, University of Otago 1996.
Sumpter, G. H. *In Search of Central Otago* Dunedin 1947.
Symons, Tod *I Know a Place* Alexandra 1978
Veitch, Betty *Clyde on the Dunstan.* Dunedin1976.
Webster, A. H. H. *Teviot Tapestry* Dunedin 1948.
Wood, J. A. *Gold Trails of Otago.* Reed Wellington 1970
Wood, S. *Mining History of Potters* Appendix 1 in Hamel 1989.

INDEX

Figures = **bold** BHF = Bald Hill Flat

Books by the same Author.
The Siren's Call, Mine Fire, Dunedin Holocaust, Coastmaster, Mountain Water and River Gold, Harbour Horror, The Golden Junction.